Best wishes

[illegible]

12/12/90

CROSSWALKS ACROSS THE UNIVERSE

A Guide to Amateur Astronomy

by

Philip J. Imbrogno

VANTAGE PRESS

New York / Washington / Atlanta
Los Angeles / Chicago

FIRST EDITION

Published by Vantage Press, Inc.
516 West 34th Street, New York, New York 10001

Manufactured in the United States of America
Standard Book Number 533-04931-8

Library of Congress Catalog Card No.: 80-54530

To Ron, Rick, Rob, Brad, and Cathy, who have endured the cold winter nights with me all these past years

CONTENTS

ILLUSTRATIONS

Figures

Photographic Plates

Tables

ACKNOWLEDGMENTS

I wish to express my thanks to the following for use of material and information which helped me in the publication of this book: David L. Chandler. *Life on Mars*. E.P. Dutton Company, 2 Park Ave., New York, New York, 1979; *Larousse Guide to Astronomy*. The Hamlyn Group; Dr. Dean R. Morelli for text information from his lab notes; Cathy Salerno, for some outstanding lunar photographs; The American Meteor Society; The Greater Hartford Astronomy Club; The American Association of Variable Star Observers; The American Association of Lunar and Planetary Observers; Milbank Observatory case files; and The National Observers of Aerial Phenomena.

INTRODUCTION

This book was written for the person who thinks he might want to try amateur astronomy as a hobby. Astronomy can be both a very rewarding and disappointing hobby. This will depend on how well prepared you are when you use your first telescope. The information that I cover will be very useful since it will tell the novice astronomer what to expect and what not to expect. Many astronomy books in the past have exaggerated what may be seen in small telescopes—the novice has the idea that he will see glowing colors and rainbow emissions in space. The universe is beautiful, but it is a subtle type of beauty one gets to appreciate more and more over the years.

These books also include photographs taken with some of the largest telescopes in the world. These photographs were taken with special equipment and very long exposures and they are breathtaking, yes, but the amateur who sees these photographs expects the objects to look like this through his own telescope and once again he or she is disappointed.

I have included a section of photography taken with amateur telescopes near or around the populated areas of the United States. Although some of the photographs may not totally do the object justice, this is about how you would see them visually in an average amateur's telescope.

I hope this book will help you reach the goals that you seek in astronomy; I have written the text in the same manner as if I was sitting down and talking to a novice astronomer about the hobby.

I call it a hobby but as you shall discover it can be more than a hobby—almost a profession to an experienced amateur astronomer. Although I myself have a degree in astronomy and have taught the subject on the East Coast in several schools, I still call myself an amateur since the word means someone who does a task for love of it and not for money.

It was also not the purpose of this book to make a great deal of money, but to inform the novice astronomer.

I look forward to hearing from my readers in exchanging astronomical projects, photographs, observations. I can be contacted at NOAP or the Milbank Observatory, both at P.O. Box 4218, Greenwich Conn., 06830.

Philip J. Imbrogno

I.

AMATEUR ASTRONOMY: MORE THAN A HOBBY

During the last decade we have seen a great surge of interest in outer space. This interest spans curiosity from intense to mild about the universe around us. A great deal of the attention that the universe is getting is due to recent scientific exploration of the solar system. Many of us have seen the fantastic pictures of the planet Jupiter sent by Voyager I and II, and this highly successful project has caused many people to visit obervatories to behold the beauty of the heavens. The movies *Star Wars* and *Close Encounters of the Third Kind* have also caused a sudden rush of new astronomy enthusiasts.

Astronomy is not a lazy man's hobby. It takes quite a bit of dedication to stay out to all hours of the night observing or photographing the little blobs of hazy light which are galaxies millions of lights years away. Many times I have braved subzero weather to observe on a clear night with no moon. An amateur astronomer must make the most of every clear night that he has since nights with excellent conditions are not common, especially if you live on the East Coast of the United States.

Many who try astronomy as a hobby are disillusioned since these people usually jump into it doing little research about the instruments to be used or the Universe itself. Many types of telescopes are on the market today and one must be very careful as to what to buy. Astronomy can be more than a hobby and if you are really interested in it you should invest in a telescope that will perform for you.

Many books on the subject suggest that you buy a small telescope and work your way up to a larger one as you grow into the hobby. I do not agree with this. These small telescopes are nothing more than expensive toys and will hinder your development. These small telescopes are bought and used a couple of times on the moon and then stored away to collect dust. Most likely it was used also to look around the sky, but all that was seen was a few stars—so the person lost interest. Losing interest was caused by two factors: poor equipment and poor knowledge of astronomy. I shall cover later in this book my recommendation about equipment for the person who thinks he might want to try amateur astronomy as a hobby.

Astronomy can be more than a hobby and amateur astronomers are needed to observe and keep track of certain celestrial events. Did you know that many discoveries were in fact made by amateur astronomers? Many comets and different phenomena in space were discovered by amateur astronomers!

There are basically two types of amateur astronomers. The first is an interested amateur who takes his telescope out perhaps once or twice a month just to look around. The second type is the serious amateur who is usually out every clear night and not only observes the sky but keeps an accurate record of the object he is studying. He plots storms on Jupiter or Mars, keeps track of the activity on the sun, photographs the distant galaxies looking for stellar eruptions.

There are a number of groups now in operation that make use of the amateur's ability to observe the heavens. These groups gather observing information on particular objects for scientific study. All amateurs are welcome to join these groups and it is a good way to make use of your talent and equipment. It is perhaps wrong to call this type of amateur astronomer an amateur since his knowledge in astronomy in the general sense is perhaps greater than most professional astronomers.

Many observatories are involved with projects that might take them years and there is quite a bit going on in the universe. How do the scientists keep track of everything? One way is working in close cooperation with various astronomical organizations. These organizations are national and worldwide and are composed totally of serious amateur astromers who are in fact the watchers and keepers of the universe. Lets take a closer look as to what you might do as a serious amateur to assist in a scientific study of astronomy.

Variable Stars

These are stars that periodically change their brightness. Studying these stars is very valuable. These stars are unstable and some of them are pulsating. An amateur astronomer can keep track of the period of the change in brightness—a sudden change in this period can mean something unusual is going on in this star. The amateur then can report this change to his group leader (if he belongs to a variable star observers organization) who will contact an observatory to verify it. This information can be extremely valuable in the study of stellar evolution (the birth, growth, and death of a star).

There are in fact thousands of variable stars and the professionals can not keep track of them all, so the job has been turned over to the serious amateur who, in this field, provides an important service for the scientists.

The Planets

There are several groups now in operation that are composed of amateur and professional astronomers that "patrol" the planets on a routine basis. These planetary "patrols" are done by group members using large telescopes (of twelve inches of more) to keep a permanent photographic record of the changing features of several of the planets.

Of course, only five of the nine known planets can be seriously studied by the amateur astronomer. They are Mercury, Venus, Mars, Jupiter and Saturn. All of the planets listed above display a big enough disk, even through a small telescope, to be studied in detail. You as a member of one of these planetary patrol organizations might study the development of a storm on the planet Jupiter from night to night, or you might study and record the shrinking or growing of Martian polar caps. On the subject of planetary observations, it would take an entire book just to cover the basics, but I shall go over the finer points later.

Many amateur astronomers have been credited with the discovery of a new feature on a planet. In the early sixties a young man was outside using his telescope one night. He turned his 'scope toward Saturn and saw a large oval bright spot in his six-inch reflector. He then made accurate notes of his finding and called the nearest observatory which confirmed his observation. This young amateur astronomer was credited with the discovery of a new marking in the cloud cover of the planet Saturn.

I myself have seen the formation of new storms hours before the professionals at the large observatories have picked it up. I would like to relate a most interesting phenomenon that I observed on the cloud cover of Jupiter in the Winter of 1980. Below is a transcript from my observation notes.

> On the twenty-seventh of February 1980 at 3h 20m UT (10:20 P.M. EST on the twenty-sixth) I observed a most unusual phenomena in the cloud belt of Jupiter. I had set up my eight-inch Schmidt-Cassegrain telescope for an evening of photographing Jupiter and Mars. I first turned the scope toward Jupiter and made a visual observation. I saw a large storm in the north-equatorial band which at once drew my attention. In the south tropical zone I saw a brownish smudge and at the time I really didn't think too much of it. I then hooked up my camera to the scope using a 9mm ocular (234X) and shot several frames each five minutes apart using SO-115 by Kodak. After I had completed the photographic session of Jupiter, I then once again went visual with a 6mm occular (350X). The area where I saw this brownish smudge before has changed and in its place was this very dark black spot with a brownish halo around it.

This was some type of eruption in the Jovian cloud belt—according to a chart that I had there were not any transits (one of Jupiter's moons throwing a shadow on to the planet) visible at that time. The area faded several hours later and was soon lost to view since it rotated around the other side of the planet and was no longer visible. The next night according to my calculations it should have been once again visible from the earth, but nothing was seen. There was no evidence whatsoever that it ever existed at all. This type of transient phenomenon in the form of eruptions is seen from time to time on the gas-giant planets. A violent event like this is however very rare. I did however document the event on film.

Many discoveries are made by accident and as in the case above I was just looking in the right place at the right time. I wonder how many others saw this anomaly. If you have I would like you to contact me.

Many people will say, "Why should I take part in a planet patrol when the space probes like Voyager and Pioneer are there? Surely I can't compete with them." This is wrong, you can. These probes survey an area in just a short time, perhaps something will occur when the probe turns its camera away from the planet. It is like saying, "Why have local weather stations when we have these weather satelites photographing the weather patterns over the entire Earth?" How many times have the weather persons reported sunny skies only to have it rain? Each area has its own weather patterns and thus small observation stations are needed throughout the country to keep track of the weather. The same is basically true with a study of the planets; small observing stations around the country are needed to support the great exploration now under way of the solar system. By studying the planets' geology, weather, and so on, we can perhaps learn more about our own planet and improve our own conditions on Earth.

In the future man will have to move out into space for natural resources and the planets may prove to be an ideal place for gathering raw materials. As we know, the moon is rich in many metals, and perhaps so are the other inner planets and Mars. We are exploring now for the future so that perhaps some day we might be able to reach these worlds for the benefit of the human race. The grounds of study and learning have to be laid first; we have to know more about these worlds, and the amateur astronomer can assist in this endeavor.

Solar Astronomy

Perhaps one of the most important functions an amateur astronomer can perform is observing the sun. Keeping track of solar activity in the form of sun spots and flares can be very useful. The sun is an average star as stars go but, to us here on Earth it is the most important star in the universe, for it keeps us alive. The sun sends out radiation in space and this radiation is in the form of not only heat and light but particles called the solar wind. The solar wind is ionized gas or "plasma" composed of atomic particles. These particles travel at a speed of over 250 miles per second. They reach the Earth anywhere from twenty-four to forty-eight hours after they leave the sun. This solar wind is ejected by disturbances on the sun such as sun spots and solar flares.

Sun spots are dark areas on the surface of the sun. They appear dark to us here on Earth because they are cooler than the surrounding solar area, and thus give off less radiation. Actually the surface temperature of a sun spot area is still thousands of degrees. The central dark area of the spot is called the *Umbra*, while the lighter brownish area that surrounds it is called the *penumbra*. The spot area is cooled by swirling magnetic fields on the sun. These magnetic vortexes swirl the area, causing it to cool. Sun spots occur in a cycle of about every eleven years and a complete cycle from maximum to maximum again is about twenty-two years. (Maximum is a period of eleven years, when the most spots are seen. Minimum is when very few spots are seen; this is also a period of eleven years, thus a complete cycle from maximum to maximum is twenty-two years).

Sun spots have an average lifetime of about a week but some have been known to last for months. Sun spots normally occur in pairs. The leading spot is always of a "northern" polarity, while the trailing spot is of a "southern" polarity. This magnetic attraction (opposites attract) keeps the sun spot pairs together as they march across the solar disk. The above is true in the northern hemisphere of the sun while in the southern hemisphere it is just the opposite. The magnetic strength of a sun spot area is over five hundred times stronger than the surrounding solar area.

Solar flares or eruptive prominences are swirls of gas ejected as far as 500,000 miles from the sun. The force needed to do this is great since the gravitional pull of the sun is very strong. The forces involved that cause these eruptions are not fully understood by science.

The radiation sent out by these disturbances has an effect on the Earth's weather. Droughts and adverse weather conditions have been linked to solar activity, and the solar wind striking the Earth's upper atmosphere causes radio blackouts. The phenomena known as aurora borealis in the northern hemisphere and aurora australis in the southern hemisphere has been linked to solar activity.

Observations of solar activity can help predict the occurrence of this phenomena to within several hours. The auroras can come in many shapes and colors and they usually last for several hours. If you live in the more extreme latitudes you have a better chance of seeing them. This is because the charged particles gather around the magnetic poles of the Earth. The luminosity arises from emissions of ionized atoms in the ionsphere. The most common colors reported are green, red or pink, and yellow.

Daily reports of solar activity can be very useful and the amateur can contribute quite a bit to solar research by joining solar study groups. A complete listing of the most popular and well-established groups are given later in this book.

Comets

The most gratifying experience in astronomy to some is the discovery of a new comet. A great many of the known comets were in fact discovered by amateur astronomers. Many amateur astronomers focus their hobby on just looking for comets. They purchase special wide angle telescopes and binoculars and scan the skies at sunset and sunrise in the hope of discovering a new comet or rediscovering an old one. Comets are "dirty snowballs" and are composed mostly of methane, water, and nitrogen. A comet consist of three parts: the nucleus which is the solid core, the coma which surrounds the nucleus and consist of frozen gases, and the tail which is matter streaming off into space for millions of miles. As a comet approaches the sun the pressure from the solar wind forces material from the coma to be ejected into space. The tail of the comet is thus formed and is always pointing away from the sun. As a comet approaches the orbit of Jupiter a small tail might start to form, as it gets closer to the sun the tail grows larger and the coma expands due to the radiation heating it. After a comet swings around the sun and begins to leave the solar system the tail disappears once again as it reaches Jupiter's orbit. To keep track of the known comets and to look for new ones is an important and exciting task for the amateur with the proper equipment.

Asteroids

Keeping track of the orbits of some of the larger asteroids is more of a job for the advanced amateur astronomer or even the professional. Most of the larger asteroids look like nothing more than stars when viewed through the telescope. The only way that one can tell them apart from the background stars is that the asteroid's position will change from night to night in relation to the stars in the area. This is because the stars for all practical purposes are at an infinite distance, but the asteroids are close to us. In fact, in comparison to the distances of the closest stars, the asteroids are "in the neighborhood." The asteroids which are closer appear to us to move faster in relation to the so-called fixed stars. The stars also have their own movements around the galaxy but being so far away it takes years to detect their movement with the most accurate equipment.

An Earthbound example of this phenomenon is the case of two aircraft flying overhead being observed by someone on the ground. One of the aircraft is a twin-engine type at an altitude of one thousand feet, the other is a military-type jet at four thousand feet. The miltary jet is flying many hundreds of miles per hour, while the twin-engine plane is flying a mere two hundred miles an hour. The observer on the ground watching both planes will see the slower plane move more quickly than the high-flying jet. This apparently faster motion of the twin-engine plane is because this aircraft is closer to the observer on the ground. As the observer looks up at the high-flying jet he sees it moving slower, with it in view well after the lower flying engine plane has disappeared from sight.

Most of the asteroids have orbits that lie between the planets Mars and Jupiter. This is about 2.8 astronomical units from the Earth. (One astronomical unit is equal to the distance from the Earth to the Sun, or about 93,000,000 miles.) The first-discovered asteroid was Ceres, found by the astronomer Piazzi in 1801. This caused a big stir in the astronomical world and soon after astronomers from all nations were engaged in asteroid searches. By the early nineteen hundreds no less than 500 were discovered. They ranged in size from 430 miles in diameter for the largest all the way down to twenty miles for the smallest. There are several thousand other objects that lie between the orbits of Mars and Jupiter, most of them the size of mere stones. These are called meteroids and we shall talk about these later.

It is most likely that keeping track of the asteroids will be more of a job than a pleasure. Their movements among the background stars cannot be drawn accurately. Their positions have to be photographed; this will be covered in the later chapter on observing.

Meteors

Observation of meteors is easy to do. For the most part all you need is your eyes, a pencil, paper, and clock and perhaps a star map. There are several in-depth meteor studies now under way that the amateur astronomer may take part in. One organization which charts and studies meteors is the American Meteor Society. This group is composed of both amateur and professional scientists who study basically two types of meteor displays.

Type-one meteors are the so-called meteor showers. These showers are associated with the paths of periodic comets. After the coma of the comet is heated by the sun and the tail is formed, the comet leaves behind it a trail of debris. This debris continues to orbit the sun and when the Earth passes through this debris in space we see a meteor shower. Observations are important to see what area of the sky the meteors are emitting from; this allows us to see where the debris lies in space. Also the intensity of the shower is recorded to determine if the shower is getting stronger or weaker as the years go by.

These particles or debris that we see during one of these showers are normally no larger than a grain of sand and are very light. Several different colors have been noted, but the most-reported colors are yellow, orange, red, and white and sometimes green. There are several different showers that occur on a yearly basis, and all come from different constellations in the sky. One of the most popular and fantastic yearly showers is the Perseid shower which is seen to come from the constellation of Perseus. It occurs on about August 12 and as many as sixty meteors may be counted each hour. These meteors usually last several seconds and burn trails that may stay visible for fifteen seconds later. Many however are fast-movers and last about a second. The slow-movers are the most spectacular to see since it takes longer for them to burn and you see them longer. Let me make it clear what I mean by a "fast-mover" and a "slow-mover."

As the Earth moves in space this debris is picked up by the Earth's gravitational field. If it picks up the debris as it's moving away from the particle we have a slow-mover or a slow-burner. If it picks it up as it's moving into the debris we see a fast-mover or a fast-burner. These meteors that we see burn up about fifty miles up and may be very bright. This would depend on the size of the particle. A particle the size of a pea will burn brighter than the full moon. In my fifteen years of astronomy I have only seen three meteors get this bright and it is a sight to behold. You can't predict when it will occur but when it does it both startles and awes you. A brilliant light in the sky appears out of nowhere, illuminating the entire ground in a bright blue-white light. This type of brilliant meteor is called a fireball. Some are even known to explode and these are called *bolides*.

I have heard reports from amateur astronomers from across the country who after seeing a bright fireball burn up have heard a buzzing or a hissing sound a short time later. This can not be verified but is interesting to note.

The American Meteor Society is the group to join if you are interested in this branch of astronomy. Many observers are needed and I'm sure your talents as an observer will be welcomed.

Another type of meteor seen is the type-two meteor. This meteor is quite different than the small grains of material that we see burning up in the upper atmosphere in the night sky. These meteors can be called true meteors since they are mostly made up of the debris of material that lie between the orbit of Jupiter and Mars. They have highly elongated or cigar-like orbits which bring them close enough to cross over the Earth's orbital plane. If one of these objects crosses over the Earth's orbital plane while the Earth is there, the object will be attracted by the Earth's gravitational field and be pulled towards the Earth. Some however, do not head straight towards the Earth but only graze the upper atmosphere. The result of this is that we see a streak of light shooting across the sky at great speed. These objects are called, in lay terms, shooting stars.

These meteors range in size from a few ounces all the way to several tons and may burn so bright that they might even be visible in the day time. While the objects are *in space* they are called meteroids, while they are seen burning up *in the atmosphere* they are called meteors. Occasionally they may be large enough to survive the violent burning caused by friction in the Earth's atmosphere. When this occurs, part of the mass will hit the ground and when this happens it is called a meteorite.

There are about three meteorites recovered each year around the world and I had the opportunity to assist in the recovery and identification of one a number of years ago.

On the twenty-seventh of July, 1977, a Mr. Robert Roth of Old Greenwich, Connecticut, called me at my office and told me that he had found a strange-looking stone. Having some geological training Mr. Roth looked at the rock and thought that it might possibly be a meteorite. I then made plans to come up to his home the next night and take a look at what he found. When I arrived I was told that Mr. Roth's six-year-old son Jeffery had found the object at a home of his friends and liked it so much that he took it home with him.

After doing some investigative work, I found out that the object was found on a beach thirty years ago by a Mr. Stan Chard in a small depression. The late Mr. Chard used to reside on an island off the Connecticut coast and would walk the beach every morning looking for shells and stones. It was one morning that, while strolling down the beach, he saw this shiny black cone-shaped stone. He knew that it wasn't there the day before since he covered that area. He then picked it up and put it in his collection where it remained until he died. Claiming his relative's possessions, a Mr. Walter Chard then took the stone to his home in Greenwich, Connecticut, where it was used as a door stop for a number of years. It was then picked up by Jeffery Roth who took the stone home with him.

The object was black in color with a very smooth surface and, to my surprise, the fusion crust was still intact. Also visible on the object were a number of depressions and cracks. The object looked as if it was subjected to intense heat. The shape was cone-like and it weighed about five-and-one-half pounds. Inside the rocky stony mass, small metallic iron and nickel filings were found imbedded.

The object was then sent to a lab in New York City where laboratory analysis showed it to be composed of the following elements: oxygen, silicon, magnesium, with traces of iron and nickel. This object was classified as a stony-type meteorite and has now taken up permanent residence in the Bruce Museum in Greenwich, Connecticut.

There are basically two types of meteorites. The first are called the *stony-type* and are similar to the Roth meteorite. These stones, if not recovered and protected soon after the fall, will weather and be unrecognizable when compared to Earthly stones.

Another type is the iron-nickel-type and are composed of these two elements. These stand out very well and are recovered quite easily since they are easy to pick out in relation to Earthly stones even after some time has passed after the fall. As an amateur astronomer, you may be called to assist in the recovery of a meteorite in your own area.

Galaxies

A scientific study of distant galaxies is an exciting experience for the amateur. There is a drawback however to the study of galaxies and that is there is very little the amateur astronomer can do. This is because the galaxies are so far away that when observed all they appear as is glowing blobs of light of various shapes. You can however with modest amateur equipment see some of the structure of the largest and closest galaxies.

There are three basic types of galaxies: first, the spirals which range in shape from pinwheels to a lens shape; second, the elliptical galaxies which are egg shaped; third, the irregular galaxies which have basically no shape but are chaotic groupings of dust, gas, stars, and most likely planets. In an average-size galaxy there are at least 100 billion stars and in the entire universe over a 100 billion galaxies.

We call our galaxy the Milky Way and the solar system is located about thirty thousand light years from the center. Our galaxy is lens shaped and is at least one hundred thousand light years from tip to tip.

When we look into the night sky towards the constellations of Lyra and Cygnus we see a faint glowing white patch of light in a narrow band that stretches from horizon to horizon. This is called the Milky Way and is the area near the central core of the galaxy.

Any amateur wishing to study galaxies must have a moderate-size 'scope and a complete knowledge of the heavens. One must know where to look to find these galaxies, since while looking for one the inexperienced amateur might pass by one without even knowing it. For the most part the galaxies are faint, but a few of them are bright enough to see even in binoculars. One of these bright galaxies is located in the constellation of Andromeda and is bright enough to be seen on a clear moonless night with the unaided eye. Later on we shall cover how to find and what you need to study these island universes that we call galaxies.

Unless the amateur astronomer has special equipment perhaps the only thing in a distant galaxy he can discover is a nova, or a supernova. In short a supernova is an exploding star and at times, before it fades, may outshine the entire galaxy that it's in. A nova is a less-violent event and are stars flaring up or going through minor explosions. Novas also can become quite bright and have been known to be seen easily in the most-distant galaxies. Several of these novas and supernovas that have been seen blazing in the distant galaxies have been discovered by amateur astronomers.

How is it done? Well, most observers look at the same galaxy at least twice a year. If something unusual is taking place the amateur will notice it since he or she knows from experience what the galaxy normally looks like. In the case of the supernova a bright star will be visible where none was before.

One case of this type of discovery was in April 1979 when an amateur astronomer, a Mr. G.E. Johnson, was observing the galaxy M-101 (these numbers are merely listings in a catalogue compiled by French astronomer Charles Messier) when he noticed a star visible in the spiral arms which seemed as bright or brighter than the entire galaxy. According to his charts no star was there previously; this was strange indeed. The discovery was then confirmed by several professional observatories from around the world as being a supernova. This galaxy lies in the constellation of Ursa Major, or the Big Dipper as it's frequently called.

This nova or supernova is the fourth such to be seen in this galaxy this century. Supernovas seem to occur at least four times a century for each galaxy. This is one of the reasons astronomers look for them in the millions of galaxies that surround our own.

I don't know of any special groups of amateur astronomers that do galactic studies as their primary function, but there are the so-called "deep-sky clubs." These groups for the most part keep their telescopes and cameras turned on the deep-sky objects. These consist of the nebulas, star clusters, and galaxies. There are many astronomy clubs and organizations around the country that offer galaxy observations as part of their regular programs. There should be several in your area.

Special Astronomical Studies

These may be termed freak events, once in a lifetime occurrences, or unusual astronomical events. Below I shall cover the various astronomical phenomena and what the amateur astronomer can do to contribute in the study of this specialized field.

Occultations

An occultation occurs when a larger, closer astronomical body moves in front of a smaller, farther one. There are many types of occultations but the moon plays a special role in the occurrence of this phenomena.

The occultation of stars by the moon is a fantastic sight. The star always enters from the eastern limb of the moon and then disappears. It then reappears from the western limb of the moon. This is because the moon actually moves just a little slower than the background stars. Observing these stellar occultations by the moon can prove to be a very worthwhile and interesting task for the amateur astronomer. Accurate timings are needed however in the recording of the events. Your first time would be when the star disappears—this disappearance will occur all at once as it approaches the eastern limb of the moon. This is because the moon has no atmosphere and no refraction phenomenon is taking place. At times however one might see slight flashes seconds before the star will disappear. This is true with mostly only bright stars and bright planets and is due to the light from the star reflecting and streaking at low angles through the lunar valleys.

Your second recorded time will be when the star appears from the western limb of the moon. Once again the appearance will be all at once without any warning. A stellar occultation by the moon can occur at any phase and time. If the occultation occurs during, let's say, first quarter, the star will vanish behind the dark part of the moon then emerge from the light portion. Any observation of this type of event, no matter the circumstances, is a rewarding experience for the amateur astronomer.

During such an event observers both professional and amateurs time the disappearance and reappearance from many locations across the country, and the data is used to match predicted times against observed times. This difference in the recording of time from a number of locations can help us to understand and study the very complex motions of the moon as it revolves around the earth. Thus the amateur astronomer can and does play a very important role in the time recording of the occultations. Many science magazines publish the recorded times of amateur observers during a major occultation (such as a very bright star being occulted by the moon). Such magazines are *Astronomy* and *Sky and Telescope*. These two excellent publications welcome the work of amateur astromers and encourage them to send in their disappearance and reappearance time of major occultations.

It is clear that one must have access to some method of obtaining accurate time during the recording of the occultation. Clocks, wristwatches and the like will not do. One source that amateurs and professionals use is a radio broadcast that gives out universal time to the second twenty-four hours a day, seven days a week. This system is called the WWV and operates on the frequencies of five, ten and fifteen megacycles in the shortwave amateur-radio band. These frequencies are available on most low-cost shortwave radios and are usually picked up strong and clear. This of course would depend on your location and the weather conditions at the time.

It must be made clear that this service gives out Universal time—that is to say, a time standard used worldwide. UT is also GMT or Greenwich Mean Time. For example, if it is 9 A.M. in New York City then in Greenwich, England, it will be 2 P.M. If the East Coast is not on Daylight Savings Time then GMT or UT will be plus five hours, if we are on Daylight Savings Time, since we put the clocks ahead an hour, GMT will be plus four hours. Thus during Daylight Savings Time 9:00 A.M. in New York City becomes 1:00 P.M. GMT or UT.

This is not the end of it. Astronomers do not use the terms A.M. or P.M. to express UT. In astronomy we use a twenty-four hour clock. Thus 1:00 A.M. is lh UT and 1:00 P.M. becomes 13hrs UT. The further west you go in the United States the more hours you must add to convert to UT. To Eastern Standard Time, as I have said, you add five hours, to Central Standard Time six, to Mountain Standard Time seven, to Pacific Standard Time eight and to AHST (Alaska, Hawaii) ten.

The procedure and timing for events are standard for observing and recording all occultation phenomena.

So far we only talked about the moon occulting the bright stars and we mentioned that it was a very interesting sight. There are many types of occultation phenomena for the amateur to observe and perhaps the most spectacular is the occultation of a bright planet by the moon. With stars, all that is seen is the star disappearing behind the moon, but with a planetary event of this kind, through the telescope, at the higher powers, we see the disk of the planet.

Venus, Mars, Jupiter, and Saturn are the most interesting bodies the moon occults. This is because they display a large disk and are very bright. These planets being occulted by the moon are even a fantastic sight to behold with the naked eye. Once again accurate times of the event must be kept. Planetary occultations by the moon are a much rarer event than stellar occultations by the moon and should not be missed. Amateur astronomy magazines will give data and forecast an event like this at least a month in advance so that you will be prepared for it.

The most interesting and beautiful planetary occultation by the moon that I ever saw involved the planet Venus. The event occurred December 26, 1978 at 11h UT (6:00 A.M. EST). I had set up my eight-inch reflector at about 5:00 A.M. to observe and photograph the event. At this time I looked toward the southeastern sky and saw the planet Venus which appeared to my naked eye as a very bright star-like object. Close by was the crescent moon. The weather at the time was clear but a little cold, but was quite comfortable as long as the winter wind did not blow. When the occultation began I was observing it through the eyepiece of my 35mm camera, which was attached to the focus of my 'scope. It was a fantastic sight. As Venus, which appeared as a sort of half moon shape with a deep white-yellow color, vanished behind the moon, I was awarded by seeing the last flicker of light from the planet sparkling through the lunar valleys. Then it was gone. The emergence from the darkened side of the moon was truly remarkable and occurred without warning at about 8:02 A.M. I was able however, to see Venus through the telescope in the morning sky all the way to 11:00 A.M.

As I thought about the event I wondered what the ancients thought many thousands of years ago if they watched a similar event. Surely they would think that the moon swallowed Venus then spit it out on the other side. Such an event would have stimulated religious thoughts and writings. But in this day and age, with all the rushing and confusion, few people paid attention to the event. This was a shame for it was a beautiful display of nature at its best. As I developed the color slides that I took I was glad that they came out well, since now I had the event recorded as it happened that day.

So far, we have talked about the moon occulting stars and planets, but lets talk about another type of occultation—the occultation of a star by a bright planet. These events are to be observed and photographed only through the telescope since the naked eye is of little value when observing an event of this type. Through the telescope we see the planet's disk and once again the best subjects are Venus, Mars, Saturn, and Jupiter.

Since these planets have an atmosphere, the disappearance will not be all at once, as with the moon. As the star approaches the limb of the planet it will start to flicker and become distorted and elongated. This is due to atmospheric refraction phenomena; the bending of light. If a planet has a very thick atmosphere like Jupiter, Venus, or Saturn, the refraction phenomena will be more intense. In cases like this the star will still be visible even when it is totally behind the planet. This is due once again to the refraction or bending of light through the planetary atmosphere. Because of this refraction phenomenon, however, the star, besides looking distorted will also flash red and green and blue. We get the same effect on Earth when on a hazy summer night we look at a star that is low on the horizon. The star rising will flash colors from red to blue and also appear distorted. The more haze, the better the effect. This occurs mostly when the star is low in the sky because we look through more atmosphere when we observe astronomical objects low in the sky. Because of this effect, astronomers will wait until the object that they want to observe is higher in the sky, since the less atmosphere we observe through the better.

By observing the refraction of stars as they approach a planetary disk we can learn much about the thickness of the planet's atmosphere. For example, if the star does not flicker much and is gone very quickly, this would indicate that the planet most likely has a thin atmosphere; if the star takes a long time to vanish and is still visible after the predicted time of the occultation we can assume that the planet has a very thick atmosphere. Stars occulted by the planet Jupiter exhibit this phenomena. However if the star vanishes all at once as if a light had been turned off, then this would mean that the planet has little or no atmosphere at all. The moon and Mercury exhibit this effect.

It was the observation of this type of occultation, that is, a planet occulting a star, that led to the discovery of the rings around the planet Uranus. Thirty-five minutes before the star was expected to be occulted by Uranus it (the star) dimmed, and this was totally unexpected. The astronomers who were doing the study were flying at an altitude of 41,000 feet over the Indian ocean. They were in NASA's Kuiper Airborne observatory which carried numerous astronomical equipment to great altitudes to escape the effects of the thick atmosphere at the surface of the Earth.

As the star dimmed the astronomers were puzzled. This dimming effect last for several seconds then it once again returned to its normal brightness, but then several minutes later it dimmed once again, then again, then again. Later it was found that the dimming of the star was caused in fact by the shadows of the newly discovered rings of Uranus as they passed over the telescope. The date of this discovery was March 10, 1977, by astronomers Jim Elliot, Ted Dunham, and Doug Mink.

Planetary Conjunctions

Since the planets revolve around the sun their positions in the sky will shift from time to time. Their distance from the Earth will determine how noticeable the chance is over a short period of time. Since the Earth is also revolving around the sun and changes its position in relation to the other planets, it adds to the drama.

When two or more planets are apparently close in the sky, that is, when the angular separation between them is small as seen from the Earth then we have a planetary conjunction. These conjunctions are strange looking if the two planets are bright planets like Jupiter and Saturn. The separation can be several degrees, but the closer or the smaller the angular distance between them the better.

I recall several years ago people were calling local police and airports reporting two UFOs hovering and then moving slowly overhead. These UFOs turned out to be nothing more than a conjunction of the planets Jupiter and Saturn. When these people were told this over TV and radio most of them would not believe it since they "never saw stars that close together before." Most of them felt as if the government was forcing the astronomers to give a phony explanation to cover up the truth. My files are full of people who have reported planets and stars as UFOs. It is the duty of the amateur astronomer to point this out if he feels that any sightings of this type of phenomena can be explained by astronomical events or phenomena.

Most of the time conjunctions will occur involving two planets, but a number of times there may be several planets in conjunction. I remember a very stunning conjunction of Venus, Saturn, and Mars which I photographed and observed in 1977. The event was breathtaking and was visible in the western sky. We (local amateurs) had traveled to an elevation of 500 feet to observe it. The night was crystal clear and it was twilight when the planets first really became noticeable. The three planets formed a triangle with Venus on the top with Mars to the right and Saturn to the left.

Normally the conjunction will last several days before the angular separation between the two planets as seen from the Earth increases and the planets move farther apart. There have been occasions when the two planets moved apart, but then back together again forming a new conjunction. This happened in 1979–1980 with Jupiter and Mars. It seems logical that if the planets move apart to break up the conjunction then they must somehow move backwards. Planets like Mars and Jupiter display a backward motion as seen from the surface of the Earth. The planets themselves are not moving backwards, but they appear to do so. This apparent backwards motion is caused by the Earth passing the planets in space.

An Earthbound example of this backward or *retrograde* motion can be given in the form of two cars moving down a highway. Car A is moving at about thirty-five miles per hour and car B at seventy. Car A however is a mile ahead of car B. Car B can see car A in the distance and from his point of view it is slowly moving forward. In a very short while car B catches up with car A and as car B passes car A—car A appears to move backward.

The planets in space also exhibit this phenomena of movement since the inner planets move faster around the sun than the outer planets. Actually the closer the planet is to the sun the faster its orbit under normal circumstances. But as you move outward in space the slower the planet seems to move, and the longer the orbit around the sun. This motion, faster or slower, is mostly apparent as seen from the Earth. The inner planets have smaller orbits around the sun than the outer planets, thus they orbit in a shorter time. For comparison, Mercury orbits the sun in eighty-eight days while Pluto takes 247 years to complete one orbit around the sun.

Getting back to the planetary conjunctions there is not much in the way of scientific study that the amateur can do in this field. A conjunction between bright planets is more or less an occasion for the amateur astronomer to enjoy its beauty. The conjunctions are usually photographed in color using color slide film. Some amateurs make large photographic prints of the event since they make very nice "art" photography. In the chapter on basic astrophotography I will explain further on how to photograph the stars, planets, sun, and moon through your telescope and how to photograph events like conjunctions with a standard 35mm SLR camera.

To try to mention all the fields of astronomical study that the amateur can contribute would be impossible in this book. I have mentioned however some of the more important examples of what you can do to make your love of astronomy more than a hobby. Amateur astronomy is one field where any person who takes it seriously enough can add to man's growing knowledge of the universe.

II

THE EQUIPMENT OF THE AMATEUR ASTRONOMER

In the last ten years a great deal of equipment has appeared on the market for the amateur astronomer to use. This equipment is for the most part very sophisticated and borders in quality on professional's tools. There are a good number of telescopes availiable from many different sources and the amateur must choose very wisely when making his first choice of astronomical instrument.

Telescopes

The telescope is the eye of the astronomer and the main tool that the amateur astronomer depends on. There are two basic types of telescopes, the reflector and the refractor. The reflector (see figure 1) uses a mirror at the bottom of an open

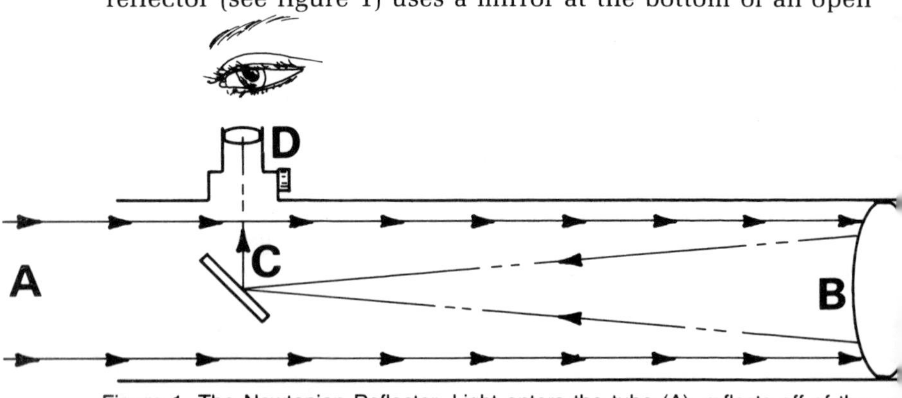

Figure 1. The Newtonian Reflector. Light enters the tube (A), reflects off of the main mirror (B), to the secondary mirror (C), to the eyepiece (D).

tube to reflect the light from a distant astronomical source. The mirror not only gathers the light, but also provides resolution of the distant source. The light is then reflected from the main mirror to another smaller mirror set at a forty-five-degree angle near the top of the tube. The light then is reflected once more into a focusing mount where it is magnified by the eyepiece. Of course all this reflecting of light through an open tube does have its problems, but the reflector has endured through the centuries and to this day it is dollar for dollar the best buy for performance.

One of these problems mentioned above is air currents in the tube, which can distort an image and cause a drop in resolution. Although this problem is encountered from time to time it does not affect an evening's work as much as you would think. Another problem of the reflector is the many times the light is reflected. This reflecting from the main mirror to the secondary at the top of the tube can itself cause a loss of resolution. This reflecting of light also causes a slight reduction of the light-gathering power of the telescope, since the reflective surfaces at best reflect about ninety percent of the light that strikes them. The rest of the light is transmitted through the mirror and absorbed.

Reflectors are also usually clumsy and hard to operate. The optics must be checked once in a while for alignment. The eyepiece is located near the top of the tube and many times you might need a stool to reach it. Because of the open tube, dust, dirt, and moisture collect on the optical surfaces and they must be cleaned at least once a year. In response to these problems, a number of different types of reflectors have been designed and are on the market today. We shall talk about them soon and how they differ from the design invented by Isaac Newton in the seventeenth century. The basic reflector described and talked about is called a Newtonian reflector after its inventor.

All telescopes have their good points and bad points, but although it seems as if the Newtonian reflector has many problems, it also has many good points. One of these good points is value. You can buy a great deal of telescope for a very low price. This means for less money you can see further into space than with other 'scopes, such as the refractor which we shall talk about later. The Newtonian reflector is called the amateurs' telescope since most amateurs own one, but as a matter of fact, the largest telescopes in the world are of this design.

Dollar for dollar the reflector is the 'scope to purchase since it will perform better for the price than any scope on the market today. The reflector is really a simple basic design and has not been much changed. The only things that improve over the years are the optics, which keep on getting better and better with the new methods of computer design. Mirrors are now accurate to a few millionths of an inch.

There are other types of reflecting telescopes available to the amateur astronomer. These scopes have attempted to solve the problems of the Newtonian reflectors, but in doing so have created other problems—problems not too severe, but there nevertheless.

The *catadioptric* telescope can be termed a modern-day reflector. Some of the problems of the Newtonian reflector have indeed been solved. Catadioptrics use a convex secondary mirror to "fold" the light into a smaller-length tube. Thus a telescope with an eighty-inch focal length using the catadioptric design will only be about twenty-four inches long. Although it is smaller (and lighter), it is still equal to a telescope with a focal length of eighty inches.

The catadioptrics come in two basic designs, the Maksutov-Cassegrain and the more-popular Schmidt-Cassegrain telescope. In either, light rays enter through a lens called the corrector plate. The corrector plate corrects the images coming in so that they will suffer minimal distortion.

(The corrector lens or plate also seals the system and protects the main mirror from dust and moisture which can ruin optical systems. This is especially important if you live near an industrial city and if your area has a great deal of pollution. Also if you live close to the ocean the salt air can cause damage to the delicate coatings of the reflector. The catadioptric, or compound telescope as it is sometimes called, has solved this problem and the optics are very nicely protected from the environment.)

After passing through the corrector plate light hits the main mirror, is reflected to a convex mirror, which reflects it back to the center of the main mirror in a very narrow cone of light. The main mirror has a hole drilled in the center of it so that

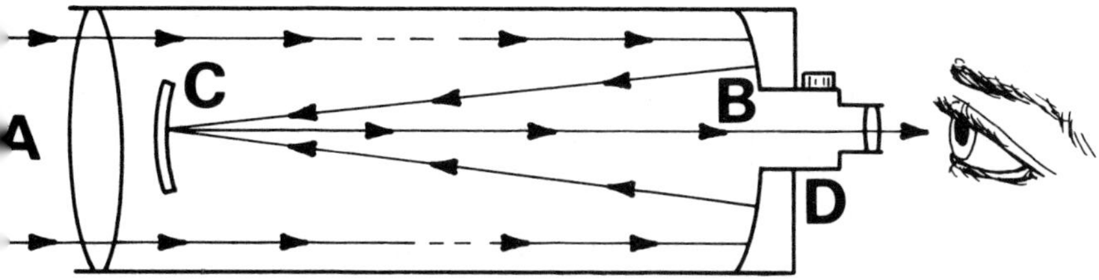

Figure 2. Schmidt-Cassegrain Catadioptric Telescope. Light enters through corrector plate (A), bounces off main mirror (B), to the convex secondary mirror (C), to the eyepiece (D).

the light can pass through to an eyepiece (see figure 2). The Maksutov-Cassegrain (see figure 3) differs from the Schmidt-Cassegrain in that the corrector plate is *aspherical*–out of round. Although more expensive to grind, the aspherical corrector plate gives a sharper image than the spherical plate of the Schmidt-Cassegrain reflector.

One of the problems of the compound telescope however is that the light is transmitted through a lens, then reflected first up then down the tube. This bouncing back and forth of light is much worse than in the Newtonian reflector. Although the factory will align the optics and, if you are careful, you may never need to adjust the optical elements, there is always that chance that for some reason the system may become misaligned. If this occurs then you have problems on your hands, it is much more difficult to align the optical system of a compound reflector than a Newtonian reflector which is much easier. Another problem is the cost and a compound telescope may run as much as $1,500 for an eight-inch model. An eight-inch New-

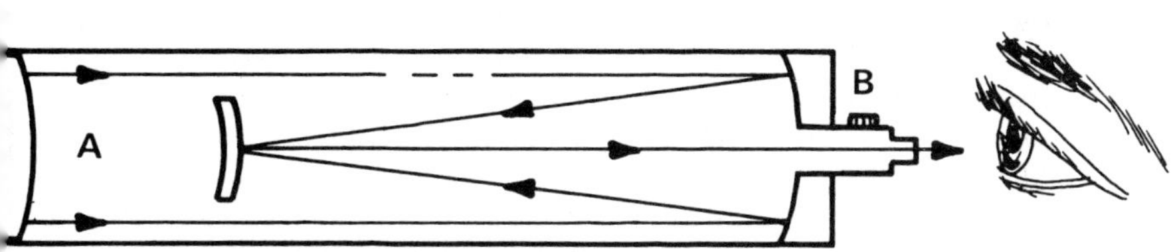

Figure 3. Maksutov-Cassegrain Catadioptric Telescope. Similar to Schmidt-Cassegrain, but with an aspherical corrector plate (A).

tonian may run any where from $500 to $800 for a professional-quality instrument.

Let's now take a look at the optical design of the compounds or catadioptric telescope. Compare the diagrams with the Newtonian ancestor. As mentioned earlier there are two basic types of the compound telescopes and both designs are very similar. The only real difference is in the construction of the front corrector lens. Since these systems do use a lens as well as mirrors, they are often called mirror-lens telescopes. However, the end product at the eyepiece is *reflected* so they are classified as reflectors.

A closing word about reflectors. First of all they are very useful for all aspects of astronomy except Sun watching (the heat built up in the telescope tube from the sun can cause damage to some of the cemented areas). Reflectors come in all sizes from a three-inch mirror amateur 'scope all the way to professional instruments. As stated earlier the size of the telescope is determined by the size of the mirror. Very rarely do you find an amateur astronomer who has a telescope larger than twelve inches.

Reflecting telescopes, especially the Newtonian type, will come in all different focal lengths. The focal length of the telescope is the distance of the point of focus from the primary mirror. This focal length determines how long the tube will be in the Newtonian reflector. To find the focal length of any mirror we simply take the f-number of the mirror* and multiply it by the mirror's diameter. For example if we have an eight-inch f/8 primary mirror then the focal length or the point were the image is formed will be sixty-four inches from the primary mirror ($8 \times 8 = 64$). The tube of this telescope will be at least this long.

In the Newtonian telescopes the f-number may vary from f/4 to f/10. Lower-f-number mirrors tend to have lower magnification and wider and brighter fields of view than higher-f-number 'scopes, make very fast photographic systems for star clusters and galaxies, and provide a fantastic visual sight for most deep-sky objects. The low-f-number telescopes provide little service for using very high powers such as are needed with lunar close ups, solar observations, and planetary obser-

*Constant for any individual 'scope, the f-number indicates the light-gathering ability of the 'scope. Note: the smaller the f-number the *more* light is transmitted to the eye.

vations. The Newtonian reflectors with the lower f-numbers are called richest-field telescopes or RFT. To qualify to be an RFT the f-number of the mirror has to be f/5 or less.

The compound telescopes are mostly f/10, but special focal reducers are on the market today which hook up between the draw tube and eyepiece and change the telescope from an f/10 to an f/5. This is great since you now have a long focal length for the planets and also a RFT for galaxies and nebula.

Most amateurs who buy a Newtonian telescope buy one with an f-number of about seven or eight. This is in between a longer focal length and an RFT, and these "mid-way" telescopes work very nicely on both deep sky objects (nebulae, star clusters, and galaxies) and the sun, moon, and planets.

Newtonian reflectors with the lower f-numbers are also a little more expensive than their longer-focal-length brothers. This is because the former mirror needs more production work at the manufacturer because the curve is deeper in the primary mirror.

I know of several amateurs who own an RFT and a long-focal-length telescope. This gives them complete coverage on all astronomical objects. This is not to say that you will need two types of telescopes to observe—an f/8 Newtonian will do nicely for all occasions.

I myself have an eight-inch Schmidt-Cassegrain made by Criterion Company in West Hartford, Connecticut. This is an f/10 scope with an eight-inch primary mirror and a total focal length of eighty inches. Since it is a compound telescope, the eighty-inch focal length is contained in a tube only twenty-four inches long. This is great for looking at the planets and what ever you want to look at. It however falls somewhat short for looking at wide star fields and clusters since they will run out of the field of view because of too much magnification. I have to remedy this with a focal reducer or as it is called a telecompressor. This when used will give me a f/5, low magnification, wide-field telescope to observe some of the larger astronomical deep sky objects.

The other main type of telescope is the *refractor*. Very few amateurs now a days own refracting telescopes since they are quite expensive. A four-inch refractor may cost more than an eight-inch reflector. Many children will buy a small refractor and many young and old amateur astronomers will also purchase them. The small "toy" refractor has a lens diameter of about 60mm (2.4 inches). Refractors use a lens located at the top of the tube to collect the light. The light is then refracted or bent down the tube to an eyepiece which causes the magnification of the image. The refracting telescope is a simple design and a very basic telescope indeed. The refractor was first used by Galileo in the 17th Century to confirm the theories of the Polish astronomer Copernicus. Refractors provide long focal lengths and a typical refractor may have an f-number of ten to fifteen. This high f-number provides excellent resolution, good contrast, and very stable sharp images. This is particularly useful in observing and photographing the planets where image stability and sharpness is needed. Unlike the reflectors, the refractors do not suffer from all that bouncing of light. The image is transmitted directly to the observer or camera. Refractors have a closed-tube system which keeps the optics free of moisture, dust, and pollution. Also, because of this sealed-tube system, no air currents are present in the tube, giving a very steady image. Let's take a look at the basic refractor and how it works. The refractors do however have their problems and one of them is cost. The refractor is usually restricted to the amateur because of the cost. The typical amateur refractor has a lens aperture of from three to six inches. This size limits the telescope as to what it can do, and will limit its performance when observing deep-sky objects which are faint and require a large aperture to see them well. Since the lens of the refractor has an f-number of around fifteen the tubes will be very long, and on a tripod mount a small refractor may look quite massive. While the limits for portability for the Newtonian reflectors are up to an eight-inch mirror at F/8 the limit for portability for the

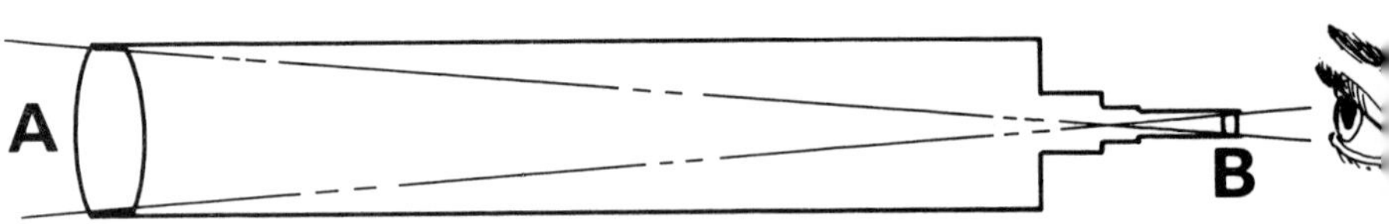

Figure 4. The Refractor Telescope. Light enters the tube through the objective lens (A) and is then refracted down the tube to the eyepiece (B).

refractor is a four-inch F/15. This means that any sizes bigger are just not made to take to different locations and should be permanently mounted.

I once had a four-inch f/15 refractor made by Unitron Company of Long Island, New York. The scope was a very good one, but taking it out and in was a problem since the scope was very massive. Refractors are also quite impressive to look at and for the most part the eyepiece is very easy to reach and many observations can be carried out sitting in a chair. In closing, refractors are basically lunar, solar, and planetary telescopes. They do also show a number of the brighter deep sky objects quite well also. Refractors, inch per inch, will outperform the reflectors. A four-inch refractor is equal in performance to at least a 6.8-inch reflector. This is only a crude guide to go on since on a number of times I have seen a refractor outperform a reflector of twice its size. In the Milbank Observatory located in Greenwich, Connecticut, this statement, became fact on several occasions. A number of us of our organization NOAP (National Observers of Aerial Phenomena, headquartered in Greenwich, Connecticut—more on NOAP will be given later is this book) were engaged in our yearly study of the planet Jupiter, and this night one of the Jovian moons was going to move across the planet's disk (a *transit*). When this occurs you can see a very dark circular shadow slowly moving across the cloud belts.

The observatory has a ten-inch f/7 reflector plus a 3.1-inch refractor attached to the tube. On the ten-inch we were using a 9mm eyepiece which gave a 190 magnification. We were using both telescopes to observe the event since there were several of us present. The 3.1 refractor is an f/15 and is from Mead Company of California. Since it was still twilight the atmosphere was still somewhat unstable and the large reflector was suffering from air currents forming in the tube. When we compared the sight in both telescopes we all agreed that the smaller refractor showed the event much more clearly than the ten-inch reflector. However, the image was much brighter in the ten. If I recall correctly, the eyepiece used in the 3.1 refractor was a 9mm one, which gave 129 power.

A small refractor outperforms a reflector of over three times its size only rarely and only with the sun, moon, and planets. Yet it does happen, especially on hazy summer nights.

What size telescope should I buy? What kind? Refractors should have a lens diameter of at least three inches. Anything less than this is nothing more than a toy and will only discourage your interest in astronomy. The three-inch refractor will show you several of the bright deep-sky objects fairly well. A three-inch refractor will give you very good views of the sun, the moon and the planets Venus, Mars, Jupiter, and Saturn. A four-inch refractor is even better, a five- or six-inch better still. The larger the aperture the better.

If you are looking for a lunar planetary scope to do serious work, I suggest the four-inch refractor. If you are in a poor "optical risk" area such as a location near the water or around an industrial city I also suggest the refractor. Reflectors are a different story, but the diameter of the mirror should be at least six inches. For the amateur astronomer a small reflector will have a primary mirror diameter from four to six inches. The four-inch mirror is very limited but the six-inch telescope will show all the planets except Neptune and Pluto. The six-inch reflector is usually the amateur's first telescope if he chooses the reflector type. This aperture is large enough to show about 70 percent of all deep sky objects fairly well on clear nights.

A typical six-inch reflector will run from $250 to $400 depending on the quality and accessories that you buy with it. This telescope is quite portable and can be moved around without too much trouble.

A moderate-sized amateur telescope is the eight-inch and a ten-inch. This aperture can resolve a great wealth of detail on the planets. These telescopes are also large enough to begin a serious study of the galaxies, since the aperture is just large enough to start to show their structure.

While an eight-inch reflector is portable to some extent, a ten-inch is not. There is quite a difference in the size jump of two inches. For example an eight-inch reflector may weigh about 125 pounds, while a ten-inch may weigh anywhere from 175 to 250 pounds. A typical eight-inch 'scope may cost $500 to $800. A ten-inch reflector may run from $1,000 to $1,500.

The large amateur telescopes run from twelve to fourteen inches. Although few amateurs have fourteen-inch telescopes, many do have the twelve- or twelve-and-one-half-inch reflector. These telescopes can photograph almost as deeply into space as much larger professional ones. They will show for example about 95 percent of all the detail on Mars and Jupiter, and show clearly and sharply just about all of the deep-sky objects. Needless to say, these telescopes are not portable. Celestron International has a fourteen-inch Schmidt-Cassegrain that they claim is portable. I have seen this scope and although it is quite compact I cannot imagine risking transporting it from sight to sight unless I had three strong men present to help me.

A twelve-inch Newtonian telescope will cost from $1,700 to $2,500. A fourteen-inch telescope might run as high as $6,000. Many colleges that have astronomy programs have a twelve- or fourteen-inch reflector or a six- or eight-inch refractor in an observatory on campus.

Look for what you need and do not go for what a company will say about how much power you can get. The useful magnification that you get will depend on your mirror or lens diameter. The rule to go by is 40X magnification per inch for reflectors and 60X per inch for refractors. Any more magnification beyond this rule will result in a blurred image with no contrast.

If you are in a bad location for astronomy, that is if your area is obstructed by trees, power lines, and so on, then it would be wise to invest in a compound telescope. A good choice would be the eight-inch Schmidt-Cassegrain which is highly portable and can be set up or broken down in under fifteen minutes.

There are also six-inch Schmidt-Cassegrains on the market today and these are very portable, but if you can spend the extra money for the eight go ahead since the six may become insufficient in a very short time. A typical eight-inch Schmidt-Cassegrain will run around $900 and weigh about fifty pounds. A six-inch Schmidt-Cassegrain might run from $500 to $750 and weigh about thirty pounds.

Telescope Mounts

Unless you buy just the tube assembly, the mount will be supplied with the telescope. The mount is an important part since when you are using the higher powers a stable, rock-solid mount is a must. If someday you might want to try your hand at astrophotography then your telescope mount *must* be solid and stable and well-balanced. Most telescope companies today provide good mounts for their 'scopes, mounts suitable for the size telescope attached to the top of them. Some companies however "undermount" their telescopes. That is, the mount is too small for the size telescope on it. This can be a real problem when you get into the medium and larger telescopes.

There are three basic types of telescope mounts used today, the Altazimuth, the German Equatorial, and the Fork mounts.

The Altazimuth

Used only on very inexpensive toy telescopes such as the 2.4-inch refractor or a 3-inch reflector. Most of the time you find refractors on them. Although they are quite stable they are not suited for astronomy and should be avoided. They provide two basic tracking motions—a circular horizontal movement which is located at the base of the mount, and a vertical movement of the telescope tube. The problem with these mounts is that in order to track or follow a celestial object you have to constantly make two movements of the mount.

The German Equatorial

This mount is one of the most common mounts in the last ten years, but now slowly losing popularity among amateurs. They provide a double-axis system, with one axis (at the angle of your latitude) pointing to the celestial pole. Since all the stars in the sky circle around the celestial pole, so will your telescope. If you have an automatic tracking system or a clock drive, the axis that points to the pole, or the polar axis as it is called, will make one complete rotation around the celestial pole once every twenty-four hours. Many of you know what the celestial pole is and where it is located. For those who do not know the celestial pole is the direction to which the Earth's polar axis points in space. In the Northern Hemisphere it lies about one degree (about two full moon diameters) from the star Polaris in the constellation of Ursa Minor. Polaris is also called the North Star and is always seen in about the same position of the sky year-round.

This polar axis provides a 360-degree movement of the telescope, this axis is also called the right ascension which follows the stars on their east to west movement across the sky.

The other axis of the equatorial mount is called the declination. This axis moves the telescope to the elevation of the object in the sky. Declination is measured in degrees with the north celestial pole at right angles to the celestial equator (zero latitude).

Since the equatorial mount is designed to revolve around the pole, it will track any object in the sky with a single movement of this axis. The declination axis need never be touched if you have a good polar alignment. This one-mode tracking of the equatorial makes it far superior for astronomy than the Altazimuth mount. The German equatorial does have its problems and they can be important for the inexperienced amateur. First of all, the mount has to be balanced with your scope. Second, at near zenith (straight up) the German equatorial has what is called blind spots. These are areas where the tube of the telescope is blocked from movement by the mount and the pier. German equatorials are common mounts with the Newtonian reflectors and refractors.

The Fork Mount

These mounts are becoming more and more popular, and many telescope producers are replacing the German equatorials with them. They are standard-type mountings with all compound telescopes and are very stable, light, and solid. They basically consist of a metal fork which holds the telescope and provides an up and down movement or declination. The bottom of the fork is connected to a plate which is connected to an axis that provides a 360-degree movement; this is the right ascension or RA axis. The fork is set on a tripod or pier at an angle to the observer's latitude. This makes it an equatorial mount and it can be used for tracking the stars with also one movement. The problems of the fork equatorial mount are minor and it is the best economical mount on the market today. It is slowly replacing all other type mounts as many companies are converting over.

I suggest you look for a telescope of the reflector type with a fork mount, but make sure the fork is massive enough to hold the weight of the telescope. If you can get a very good deal on a German equatorial by all means buy it since they are also good mounts and still widely in use. The equatorial mount should also be massive enough to hold the telescope.

Many telescope companies will give the specifications of their mounts and list the shaft diameter. A six-inch reflector of the Newtonian type shoud have shafts no less than 1.5 inches in diameter. An eight-inch 'scope should have a 1.5-inch base-shaft also, but a 1.75-inch one will work better and might be needed for long-exposure photography. A ten-inch telescope should have a 1.75-inch shaft, or more; a twelve, a two-inch shaft and a fourteen a three- to four-inch shaft.

Also, the pier that the equatorial mount is attached to for a medium scope should have a diameter of four inches, while a large 'scope should have a pier with a six-inch diameter. Mounts can be very expensive and it is better to buy them as a package deal with your telescope. Most companies offer this and it is a good saver. Most German equatorials that have refractors and reflectors on them usually have the telescope mounted in a cradle. This is a very good feature and should be looked for in a mount. This cradle enables you to slide the telescope up and down the mount and rotate it at least 200 degrees. This adjustment is quite a help since it makes balancing the 'scope much easier.

Clock Drive

The clock drive or motor drive, as mentioned earlier, is an automatic tracking system that makes the telescope move with the rotation of the Earth. Most telescopes today of the medium to large type come already with a clock drive installed on them. The clock drive is attached to the polar axis of the equatorial mount. It consist of a large gear on the axis which fits into the unit's housing. The unit is powered by a motor and the combination of the reduction of the gears makes the mount turn one complete turn every twenty-four hours. The clock drive is a must for observing the planets under high power since without the automatic tracking the planet would stay only about thirty seconds in the field of view at 300X. This would mean that you would have to move the tube and when you do this at this power you will cause a great deal of vibration which will hinder any serious study of the visible features.

A clock drive also comes in handy when you are having a "star party." A star party is when a group of people come over to your observatory or observing location and take in an evening of the beauty of the universe. During one of these star parties the observer at the eyepiece will change quite frequently. Since without a drive the image may wander out of the field of view before the third guest, and this person will probably not know how to move the axis to find it once again, a clock driven telescope will keep the image in the field of view for quite a long time and make for a more pleasant observing session.

All clock drives are not created equal. Some are better than others. If you are interested in getting started with astrophotography a good clock drive (as well as a great deal of other equipment) is a must. Some of the driving systems that are supplied on the "economy" line telescopes are very poor, and the gears are not of the best workmanship. This is why the price of the system is so low and you may think that you are getting a deal, but you're not. They are cheaply manufactured. The driving system is not smooth and the entire system contains a great deal of backlash (when the object is centered there is play between the worm gear and the main spur gear that causes the object to take some time before it is tracked by the system). Backlash is a great problem for the cheaper drive systems and can cause great frustration when observing, since it may take as long as one minute for the drive to engage and track the object that you are observing.

You can tell a good drive from a bad drive by reading the manufacturer's specifications. For visual tracking the main gear should be no less than half the objective's diameter. For example, an eight-inch telescope should have a four-inch gear. If you are doing photography, then the drive gear should be at least three-fourths the size of the objective, for smoother operation and accuracy the size of the objective. So for an astrophotography tracking system, the main clock drive gear for an eight-inch reflector should be from seven to eight inches in diameter. In general the bigger the gear the better the tracking system.

The "worm" in which the main gear fits into should be one-inch steel or brass. Brass is okay but steel is a much better system and will provide smoother tracking.

The gears also should be precise and made for astronomy with the required number of teeth for the size gear it is. Some companies will supply a plain old machine gear and this is not very good for astronomy.

The motor of the clock drive is also an important part of the system. For a six-inch or less, the power of the motor should be the minimum of thirty torque ounces. For anything larger, a 100-torque-ounce motor will be needed.

Some of the larger telescopes and the Schmidt-Cassegrains have two motors that drive the telescope—this system is very good indeed and the tracking is very smooth and accurate. A clock drive should engage and track an image no more than ten seconds after the object is centered in the field of view. I think the best drive system that I ever saw was on my four-inch Unitron refractor. The system consisted of a four-inch main gear with over two hundred teeth and a steel worm. The tracking was smooth and engaged without hesitation. There was no backlash whatsoever in this unit and it would track for hours.

A clock drive system however will track only as well as your alignment to the celestial pole. If that old polar axis is far from the pole you will get tracking errors.

At the Milbank Observatory we have an eight-inch drive gear on the polar axis of our ten-inch reflector. This driving system was custom made by Edward Byers of California. Mr. Byers produces some of the best clock drive gears for astronomical use. We bought a system from Mr. Byers after we took our clock drive casing apart and found a four-inch gear driving that massive ten-inch Newtonian reflector. The gear supplied by the manufacturers was not even an astronomical gear but a machine gear similar to the ones used in cars.

This is what you have to watch out for because many companies will cut corners in their products just to lower the price to make it more appealing to the consumer. I have been into telescope buying for over ten years and own quite a few 'scopes, and believe me I learned from experience, the hard way. There are however a good number of companies that only supply the best equipment at moderate prices. A listing of the better telescope suppliers in my opinion is located towards the end of this book.

Most clock drives are set at a sidereal rate. This means they move the same speed as the stars moving across the sky. But the moon, sun and comets move at a slightly different rate of speed. The moon moves slower, and a comet will move faster than the background stars. Although you will not notice the effect if you are observing with your eye only through the telescope, it does make a difference for photography.

For these different tracking rates we can add in conjunction to the drive what is called a variable speed control. This is a electronic unit that is hooked between the power source and the drive motor. It has a number of tracking rates or speeds for sidereal, lunar or solar. The unit is designed also with hand controls with a fast-slow dial to change the speed of the drive to even track the fast-moving comets.

A unit as described above will cost between $100 to $200 depending on the quality. Variable speed controls are a necessity for long exposure astrophotography. Normal drives without the units are set at a sidereal tracking rate and will be all right for visual tracking and short-exposure photography of the sun, moon and planets. Some of the more expensive telescopes with drive systems will include a variable speed control with the basic price. This may not be normally true with the Newtonian reflectors, but the Schmidt-Cassegrains do come with their own variable speed control. This is because the compound telescopes, especially the Schmidt-Cassegrains, have been specially designed for photography and the tracking speed units are carefully designed and tuned for them.

Setting Circles

Every celestial object in the sky has a designated coordinate and there are two coordinates for every object. These two are expressed in degrees, hours, and minutes. One circle is located on the declination axis and is graded in degrees. This marks the angle of elevation from the north celestial pole, which is ninety degrees, to the horizon, which is zero degrees. The other circle is located on the RA axis and is graded in hours and minutes from 0 hours to 24 hours. This marks the movement of the object from east to west.

It is beyond the scope of this book to fully explain the use of the setting circles. Telescopes that have them come with complete information on how to set them and find the distant galaxies and nebulas with them. Some deep-sky objects are bright, and with a good star map that marks their position are quite easy to find. Some however are very faint and you have to compute their coordinates on the setting circles, which also moves the telescope in the direction of the object you seek. Once you know how to use them they really come in handy. Without them if you are looking for a faint unmarked galaxy you may take a long time or never even find it at all. I know an amateur astronomer who never uses the circles and at times it may take him over a half hour to find one of the fainter galaxies. With the circles however I have found the same galaxy in a matter of minutes. The circles on your telescope should be at least four inches in diameter and should be made of metal. Some of the smaller telescopes might have fiberboard circles to reduce the cost of the telescope. This is not acceptable since during wet nights with dewing the circles might decay and warp over a period of time. This will greatly reduce their accuracy.

Finders and Guide 'Scopes

A finder or guide 'scope is a small wide-angle low-powered refractor-type telescope which is hooked on the side of the main telescope. The purpose of the finder is to help you locate objects. It is supplied with a crosshair reticle eyepiece and when the object is in the center of the crosshair it will be centered in the main scope.

These 'scopes are mainly used to act as a rough locater, and since they are telescopes in their own right you may be able to see what you are looking for in the finder 'scope. Many objects like the planets, moon and some of the brighter deep-sky objects can be easily seen in the finder. Once you have your object in the finder crosshairs it will then be visible in your main telescope which provides a much better view of it.

The larger the telescope, the bigger the finder that will be supplied with it. A four- or six-inch telescope will most likely have a 6 x 30 mm finder. This means that the finder is six power with a 30 millimeter (a little over an inch) objective. An eight-inch or larger telescope will come with a 12 x 50 finder, which is a great finder with a very wide field of view. A 12 × 50 finder is about the best finder for any 'scope since it will also show the fainter objects that you might be looking for. This makes it much easier to locate an object and to observe it with the main scope.

Any finder less than 6 × 30 or 4 × 30 is not worth using, and you might be better off just looking through a hollow tube. Finderscopes should always be of the refractor type since this is the best type of system for this function. Your finder 'scope will be completely adjustable to properly collimate it with your main telescope. Finders designed for the big twelve- and fourteen-inch 'scopes are generally 25 × 60 mm. These finders are actually small telescopes and will be much too large for small and medium reflectors. However, they do make good guide telescopes for long exposure astrophotography. Guide telescopes are used as a second telescope hooked on the main scope to check on the object you are photographing. This "guiding" as it is called is to make sure your drive unit is tracking properly. A guide 'scope is needed for long-term astrophotography since with the camera attached to the main scope you won't be able to see the object that you are photographing through it when the shutter is open. Guide 'scopes should be of the refractor type and have an objective of at least 2.4 inches. Some larger telescopes have very large guide 'scope. For example the six-inch Unitron refractor has a four-inch refractor guide 'scope, with a 2.4-inch finder. Guide 'scopes can also be used as a second telescope. As stated earlier, the ten-inch reflector at the Milbank Observatory has a 3.1-inch guide 'scope. We also use this 'scope for observing and photographing the sun. This is where the 'scope comes in handy, since the large reflector is not suited for solar observation. Some of the more expensive telescopes come with guide 'scopes, but many amateurs will purchase one later on and attach it to their 'scope. A typical guide 'scope of 2.4 inches will run about $125. A three-inch will cost about $180. The size of your mount will determine what size guide 'scope you can have. A mount designed for a six-inch reflector will just about hold a 2.4-inch guide 'scope, but it will never hold a 3.1-inch. An eight-inch reflector mount will hold a 2.4-inch very well, but it might have just a little trouble with the 3.1-inch. It will take a ten- or 12-inch reflector mount to hold a 3.1-inch 'scope without any balancing trouble.

A mounting system designed for a four-inch refractor such as the mount supplied by Unitron Company will hold the 3.1-inch guide along with the 4-inch main 'scope quite well. The Unitron mounts are designed primarily for refractors and are very massive and stable. If you are more interested in the products by Unitron Company you may write them for further information. The address as well as those of a number of other outstanding telescope suppliers can be found at the back section of this book.

Guide 'scopes can be used with refractors and Newtonian reflectors only. This it because the tubes are long and another 'scope of smaller size will fit nicely on them. Guide 'scopes can not be used with compound telescopes since the tubes are too short. The compound telescopes use a different system called "off-axis" viewing. In this a tube with a prism in it is hooked between the main 'scope and the camera. The prism deflects the image of a star which is off to the side of the field and reflects the light from the star up an elbow tube into an eyepiece. Off-axis guiding is becoming more popular and is even starting to be used with Newtonians and refractor systems. The photographs of deep-sky objects found in the photographic section of this book were taken with my eight-inch Schmidt-Cassegrain using an off-axis-type guider. Most guiding systems come with crosshair eyepieces so that the object may be kept centered during photography. Many of these systems have illuminated crosshairs with adjustable intensity which makes the guiding much easier on the eyes.

Eyepieces

The eyepiece is the last stop of the image through the optical system before it enters your eye. Actually eyepieces are more important than the amateur astronomer realizes. A good-quality eyepiece will insure a sharp image and high resolution even at the higher powers. It is the job of the eyepiece to magnify the image from the main objective. The focal length of any eyepiece is given in millimeters, and the shorter the focal length the more magnification it will give. Thus a 9mm eyepiece will provide more magnification than a 12mm when used with your telescope. The total magnification will depend not only on the focal length of your eyepiece, but also the focal length of your telescope. An eight-inch f/8 telescope has a focal length of sixty-four inches. To find the magnification of any eyepiece we must first change the sixty-four inches into millimeters. This can be done by multiplying by twenty-five (since there are about 25mm per inch). Multiplying by the focal length equals 1600mm. We then take the 1600mm and divide it by the focal length of the eyepiece. Thus a 9mm eyepiece will give 177X when used with an eight-inch f/8 (1600 divided by 9 equals 177); a 25mm will give 64X, and so on.

There are five basic eyepieces available to the amateur astronomer, but only three of them are even worth using. The first and perhaps the cheapest is the Huyghenian. It is nothing more than two plano-convex lenses spaced a small distance apart. The field of view is very narrow—less that thirty-five degrees. They can be found supplied with the cheaper telescopes and are good for general viewing, but are useless at the higher powers.

Another type is called the Ramsden and is composed of two plano-convex crown-glass lenses facing each other. This type of eyepiece is often found on finders and guidescopes. The field of view is about thirty-five degrees can be used at moderate to high powers. The Ramsden though better for astronomy than the Huyghenian type is still not good for serious astronomical work. It is adequate however for finder and guide 'scopes and should be used only for such. Ramsdens are also inexpensive and can also be used as a junk eyepiece when you don't want to risk a better one.

The most popular amateur eyepiece is called the Kellner and this is for two reasons: performance and cost. It is perhaps the best eyepiece for the price. A Kellner has a very large field of view—about fifty to sixty degrees, is color corrected, and performs remarkably well under high magnification. They basically consist of a crown achromatic field lens spaced a short distance from a cemented doublet lens. I have several Kellners in my collection and they all perform very well and give excellent images. A typical Kellner will cost from $15 to $20 and is worth the investment. They also function quite well for high resolution planetary and lunar photography.

The orthoscopic eyepieces are expensive high quality eyepieces. They are found as standard equipment with some of the larger telescopes. They consist of a triple color-corrected field lens and a crown glass eye lens. They are somewhat expensive super-Kellners and run as high as $40 each. They are great for high power use and the images are crystal clear even at 400X. I have one 9mm orthoscopic eyepiece in my collection which I use for high-power planetary and lunar work. The orthoscopic is becoming a popular eyepiece with amateur astronomers who are into serious scientific study of the planets.

The last type of eyepiece is a special wide-angle type called the Erfle. These are to be used with richest-field telescopes and provide up to seventy degrees field of view. The Erfle is designed for low- and medium-power use observing star clusters and star fields, bright nebulae, and the fainter larger galaxies. An Erfle eyepiece consists of flint and crown lenses spaced a distance apart from a flint crown and a flint eye lens. A Erfle is very expensive and may cost somewhere between $50 to $80. I myself do not own one since I feel a low-powered Kellner such as a 40mm or 25mm will show you everything you would want to see at that level. I have looked through an Erfle with an eight-inch f/4 RFT and I must admit the view is spectacular. What impressed me the most I think was the field of view rather the celestial object that I was observing at the time.

There are new types of designs out with different names, but all are the same optical systems described above, just arranged in a number of different combinations. I suggest that you have at least three eyepieces with your telescope. These three should include a low-power 25mm Kellner, a medium-power 12mm Kellner, and one high-power 9mm or 6mm orthoscopic. You might also want to buy a Barlow lens. This is a negative lens in a tube which fits into the focuser of your telescope. When an eyepiece is put into the top of the tube it will double the magnification of any eyepiece used with it. Thus three eyepieces then become like six. Barlows are great when observing the planets since they help you get the little extra higher power that you might need. In many cases they also improve the performance of the eyepiece. A Barlow lens is not usually supplied with a telescope as a package deal, but a good Barlow lens will cost only about $20 and is well worth the cost.

Basic Filters

Filters are used in astronomy to help bring out contrast in the object that you are observing. They also block out other types of unwanted light or transmit the wavelength that you want. Astronomical objects may emit one part of the spectrum more than another and using a filter which is sensitive to this wavelength will enable you to see the object better.

Filters for astronomy come in all colors from red to blue or violet and they can be very useful for photography. The planet Jupiter for example reflects a great deal of yellow light and by using a yellow filter to observe it we are able to see the detail in the cloud belts more clearly.

Filters screw right into the base part of the eyepiece and are interchangeable. Filters for astronomy are inexpensive and should cost no more than $10 apiece. They are composed of fine optical glass and should be protected at all times from dust and dirt. I think there are about twenty or thirty filters on the market today. All have different shades of the basic colors. You only need four filters to have the spectrum covered—any more are almost useless and may be just a gimmick to get you to spend more money on equipment.

The four basic filters are listed below and information is given about their use in astronomy.

Yellow

This may come in a dark yellow or a light yellow. I feel the light yellow filter is the better one. This is so because the dark yellow one will cut down on the amount of light reaching your eye, and this may be a problem at high powers when you want all the light you can get. The light yellow filter gives nice bright yellow-white images and is extremely useful when used to observe the planets Jupiter and Saturn. The light yellow filter helps bring out the equatorial bands on Jupiter. Since these cloud bands are brown the filter provides a lighter background to make the bands stand out darkly. It is also great for viewing the formation of storms in the Jovian equatorial area which appear as dark spots and lines. With an eight-inch reflector at about 250X you may be able to count six or seven Jovian cloud bands. When the filter is added you may see as many as twelve, as well as a greath wealth of detail that you could not see clearly before without the filter. The light yellow filter also brings out the Saturnian cloud bands which are much fainter than Jupiter's cloud belts. I will also bring out clearly the shadow of the ring stretching across the disk of the planet. The ring shadow of Saturn may look greyish without the filter, but with the filter it is very prominent and blackish-grey in color.

Dust storms on the planet Mars are yellowish in color and they are usually very hard to see. This is because the reddish color of Mars masks them. By using a yellow filter we increase the proportion of yellow light coming to the eye. In this way we subdue, so to speak, some of the redness of the Martian surface and transmit more of the yellow light. This allows the amateur astronomer to view the dust storm or cloud more clearly.

I have found also that a yellow filter helps you to see the polar caps, especially the southern one, which is hard to see most of the time. This is because during a favorable opposition the southern hemisphere of Mars is pointed away from the Earth.

The combination of a solar filter and a yellow filter will also help you to observe sun spots better. This is a special type of procedure and will be covered later since solar observing can cause blindness if safety precautions are not taken.

Green

A green filter works in basically the same way as a yellow filter. Green is nothing more than yellow and blue together. Green filters will show a considerable amount of detail in the equatorial regions of Jupiter and Saturn. A filter of this color seems to be superior to yellow in studying this region, but the yellow will show cloud detail much better over other parts of these two planets.

Many amateurs have reported that while using a green filter on the planet Uranus they have seen what appears to be a very faint equatorial band. Of course, these amateurs were using reflecting telescopes larger than ten inches. I thought that at one time I may have also seen a thin faint equatorial band on Uranus. At the time I was using a ten-inch reflector with a 6mm Kellner eyepiece (300X) and a green filter. I recorded the event in my notes but could not say for sure if it was really there since it faded in and out. I also noted it several nights later, but was still unable to confirm it. Since I no longer use a ten-inch, but an eight I have not seen it since. Still reports still come from a small number of amateurs who report this thin equatorial band in the cloud cover of Uranus. Green filters also work very well on the sun but as stated above you need another type of filter in conjunction with the visual filters to observe the Sun.

In theory a green filter should also work on Neptune, but this planet is so far out in space that only in the larger 'scopes does it present a sizable disk. A serious study of this planet is out of reach for most amateur 'scopes. Even the larger 'scopes show only a small, faint, bluish disk.

Red

Red is another basic filter. There are two types—light red and dark red. I find that the dark red makes high-power planetary images too dark, the light red is better but also is a little too dark. The filter in this color range that works very well for visual observation when a reddish filter is needed is orange. An orange filter will show a much, much brighter image when used. This is not to say that the red filters are useless—they have an important function in astrophotography. For visual use however, I suggest you use an orange. An orange works very well for Mars. It will show a great amount of detail by adding contrast to the Martian surface. This helps bring out the green-black Martian surface markings. If you want to see surface markings on Mars you will have to observe with a light red or orange filter. The red or orange filter will not show the polar caps well, however, and some amateurs agree that no filteration does make the polar caps appear brighter. You will have to experiment yourself with different filteration when observing Mars to see what works better for you.

The orange or red filter will also help penetrate the haze in our atmosphere to provide a clearer image (sometimes). This filter will also cut down the amount of light pollution (stray street light, and so on) and provide a darker background sky.

Blue

Blue filteration is seldom used. It might help you see certain areas in the Jovian cloud belt, and it might also help you see the high-altitude Martian ice clouds. Otherwise it is primarily used in astrophotography to help bring out reflection nebulae.

We have talked about the visual effects of filters. Filters visually work well with the planets to help you bring out detail by creating contrast. Filters also work well on the sun and can be used to help tone down the bright lunar light when the moon is viewed at the lower powers. The moon is so bright through the telescope at the lower powers that after observing you experience a kind of snow blindness. It is the same effect as when you walk into a dark room after walking through the snow on a bright sunny day. Everything looks dark for a while until your eyes once again become dark adapted. The moon also has this effect on your eyes after an observation. Photographically, filters provide somewhat of a different effect. They help bring out the finer detail in planetary photographs. Using a color filter with black-and-white films works best since the combination of the color film and the filter will provide a false color in your shots.

Visually, color filters offer little value when observing

deep-sky objects, but the right filter with the right film can bring out detail in the galaxies and nebulae that the eye can not see. For example, a red filter used with a red-sensitive film such as Kodak 103ae will allow you to photograph the hydrogen alpha emisson of nebulae. It will also help you photograph detail in galaxies which emit red light. This combination will also cut out quite a bit of the light pollution and transmit only the light from the celestial source. Without filtration in exposures of an hour or greater, the light pollution can get so bad that it may start to wash out the object that you are photographing.

There are several ways to cut down on the light pollution problem. We shall talk about a new method later which is perhaps the greatest advancement in astronomy in the last twenty years. This new method now helps the amateur astronomer view the distant faint nebulae clearly and brightly from light polluted urban skies.

The amateur astronomer should have at least three filters, a yellow, light red or orange, and blue or green. They will help a great deal especially in the visual observation of the planets.

Nebula Filters

This is an interference filter which blocks out stray light from mercury vapor lamps (such as found on the average American highway and street) it also blocks out high and lower pressure sodium light, (the orange-pink lights which are also called anticrime lights—they are called anti-astronomy lights by many amateurs). The filter will also block out the natural air glow of the sky.

The filter transmits the types of light that the nebulae emit, resulting in a much darker background sky and brighter, more detailed nebulae.

In a way, it provides the luxury of dark-sky country-type observing from your own backyard. The nebula filters work. And they can give you a good look at faint nebula, revealing fine detail in their structure. Without it you may see a faint blob of hazy light. One example of this is the owl nebula in Ursa Major. This nebula is a planetary type and a very faint object to observe. Observing with the nebula filter will make the difference in seeing it and not seeing it at all.

With my equipment I have a nebula filter made by Meade Instruments Company. This filter is in a standard cell and screws into a number of eyepieces of different makes. I use this filter whenever photographing or viewing nebulae with my Dynamax eight-inch Schmidt-Cassegrain system.

One word of advice: these filters are meant for only looking at and photographing the nebulae. Nebulae are gas in space and can exist as faint wisps or big bright intense clouds. The nebula filters do not work well on galaxies and star clusters. Actually these objects look worse through these filters since they do not emit the light that the filter transmits. If photographing a deep sky nebula with one of these filters, black and white film should be used. Color film will give you a very orange-red image with a greenish background. A very good film for photographing the nebulae is Kodak 103af. This film is very fast and is sensitive to all areas in the visible spectrum.

One very clear night I decided to take the telescope out and do some looking around the galaxy. I set the setting circles to the coordinates of M-27, a planetary nebula in Vulpecula and then focused the image with a 25mm Kellner eyepiece which gives about 70X in my eight-inch. I looked at the object first without the nebula filter and it looked quite good. I could make out the shape clearly as an hourglass shape. M-27 is also called the "dumbbell" nebula (this is because it look somewhat like barbells that weightlifters use) or the hourglass nebula. As I watched the glowing gas it looked green-white and I thought to myself that even if I looked at it with the nebula filter there was no way it could be seen better. I was wrong. I then took out the eyepiece and screwed the filter into the bottom part of the cell and put the eyepiece back into the focuser of the telescope. I eagerly focused and when the image was clear I was so shocked I almost fell off my observing chair. Here in front of me was the most spectacular cloud of gas in space that I had ever seen in my entire life. I saw a very dense greenish hourglass cloud, and I also saw the finer filaments which stretched around the object. M-27 now looked almost oval or circular and I could trace very faint wisps of gas quite far from the object, stretching in all directions.

I then turned the telescope away to observe some other bright nebulae. All appeared very dynamic indeed with the filter. The background sky which was a dark grey without the filter was jet black with it. As I looked at the images of the nebulae that I observed that night I compared them with photographs taken of the same objects with some of the largest telescopes in the world. To my surprise there wasn't that much difference. I could see just about everything that these observatory photographs showed and this was rewarding to me. I was somewhat skeptical about these filters when I first ordered one, but I was very gratified when I began to see these objects as I have never seen them before.

The nebula filter can be found in any catalog that sells astronomical equipment. At one time these filters were restricted only to companies that provided the amateur and professional with top-quality instruments. In the last two years they have become quite common and easy to find. They are still very expensive and may run anywhere from $80 to $200. Since there are so many on the market each one has slightly different specifications and will block out more of one type of light pollution than another, I suggest that you look your area over and find out the type of lights that are giving you the worst trouble. Determine whether they are mercury vapor, sodium, or whatever, then buy a filter which will work best blockingout this type of light. The nebula filter is an investment which the deep-space observer will not ever regret.

Solar Filters

These are filters which when used in conjunction with your telescope will provide safe solar observing. The only safe type is called the full aperture filter. This is a very large filter in a cell which fits over the front of your tube. In this way the sun's light and heat is reduced before it reaches the objective or mirror. There are cheap so-called sun filters that are supplied with toy department telescopes. These "sun filters" are nothing more than a darkened piece of welders glass in a filter cell. They screw into the eyepiece and reduce the sun's light and heat before it reaches the eye. These filters are not safe and should be banned and taken off the market at once. The problem is that when the light reaches the filter it is already magnified a hundred-fold and if the filter cracks while the observer is looking at the sun, the light of the sun will come through the cracked filter blinding the observer for life.

Once again the only safe type of filter to use on the sun is one that stops the heat from the sun before it hits the objective lens or mirror. The full-aperture filter is the best way that I know of.

At one time these filters cost hundreds of dollars. This no longer the is case, thanks to Roger W. Tuthill of Tuthill Astronomical Equipment, Inc. Mr. Tuthill has designed and perfected a full-aperture solar filter which he calls Solar Skreen. Solar Skreen provides high resolution of sun spots and is totally safe for observing the sun. The price of this filter will vary depending on the aperture of your telescope tube, but it is completely affordable and within the range of the amateur astronomer.

I have used the Solar Skreen with a three-inch and four-inch refractor and I have seen breathtaking views of the solar surface. Because of the material, the image may look a little blue. This can be corrected by adding a light yellow filter to the eyepiece. This will provide a yellow-white solar disk and will look quite natural. For solar photography the Solar Skreen works well and will provide high-resoultion photographs of sun spot complexes. I suggest however that a yellow filter be used with the eyepiece if color film is used. Black and white films need no color filters to achieve a proper image balance.

The Solar Projection Screen

This is another safe way of observing the sun. This is a whole screen which is held in front of the eyepiece. The sun's image is projected on to the screen. This method is great for group viewing and will show a good amount of solar activity. However, it has two basic drawbacks, the first and most serious of which is that the sun's heat and light is not stopped before it enters the tube. You get very serious heat buildup in the optical system and if you are using a reflector the result could be damage to your optical system which might result in the ruining of your reflector. No reflector should be used on the sun unless you are using a full aperture filter such as Tuthill's Solar Skreen. The second drawback of projection is that it is only good at low powers. At the higher powers the image becomes blurred and dark and becomes quite worthless for a serious study of the sun. Refractors usually come with solar projection screens since they are best suited for this type of observing. It is all right to use your refractor for solar projection providing your optical collimation is correct and that it be done for short observing periods—fifteen to twenty minutes.

In this chapter I have covered the basic equipment used in amateur astronomy today. There is, however, a great deal of support equipment that I have not covered, since I feel most of them are nothing more than gimmicks and really not important for general observing programs.

III

OBSERVING THE SOLAR SYSTEM

It is the purpose of this chapter to discuss observing objects in the solar system. I will go over what you will be able to see and what you will not be able to see with amateur-size telescopes. Actual observing may be a little difficult at first to the unexperienced amateur astronomer and you may not see everything exactly as I cover it in this chapter. In time, with a great deal of observing hours behind you, your eye will develop a sensitivity for picking out fine detail in the planets, moon, and sun. Many times I have had friends over to my observatory who never looked through a telescope before at the planets. When I say to them as they are observing their turn, "Do you see the storm developing?" Do you see so and so, they would turn to me and smile and say "no." Yet when I look or another experienced observer present looks we see these details clearly. Many times people think that I'm putting them on or that I'm even having hallucinations. I don't know for sure the true answer, but I do know that I can see more detail with the eye I use when at the telescope than I can with my other eye. Perhaps as you use one eye more and more it becomes sensitive to picking out more finer detail—perhaps even the cells develop more. Since we will be talking about seeing markings and faint cloud bands on the planets also it is very important that the eye be trained to pick up very minute features.

Observing the Sun

I guess a good place to start in the solar system is the sun. The sun's apparent size in the sky is quite large and it can be studied even with the smaller amateur telescopes.

First let me make a remark once again on how dangerous the heat of the sun is at the focal point of an unfiltered telescope. An unfiltered beam of light from the sun through a four-inch refractor will ignite a piece of paper in an instant. It will even burn and set fire to wood held at the focus and can of course burn the eye and cause blindness. In my early years of astronomy I had one of those toy refracting department store telescopes and one day I decided to look at the sun. The telescope came with one of those so-called "solar filters." I turned the 'scope towards the sun and began observing. After about a minute I heard a crack and just as I moved away from the eyepiece a bright beam of light came flashing from the telescope. The filter had cracked. I did however as I was backing away catch a small section of the beam in my eye and I was partially blinded in the eye for two weeks. Luckily my vision returned to normal and I got back all the sight in that eye, but I learned a lesson I will never forget.

We shall cover observing the sun with a full-aperture filter only. To observe the sun you will need at least a three-inch refractor or a six-inch reflector. This is not to say that a 2.4-inch refractor or a four-inch reflector will not show you sunspots, but a serious study or a good view can not be obtained with these small instruments. A three-inch refractor might show the surface of the sun better than the reflector since it provides more contrast and sharper images.

With the three-inch or six-inch, start out at low powers—perhaps 30–50X. A good way to locate the sun in the telescope since you will not be able to use your finder is to line up your 'scope visually with the sun. This can be done by pointing the telescope toward the sun and lining up the shadow the telescope will cast. When the shadow of the tube is perpendicular to the sun then it is approximately pointing at the sun. Now look into the eyepiece (of course with the full-aperture filter on the tube). You may not see the sun, but you may see the sun's light. Slowly move the tube around in a circular motion. The sun will come into view.

At about 50X the sun's disk will be large and you will see the sunspots on the bright solar disk. Remember the solar filter is blocking out over 99 percent of the heat and light.

I think you will be surprised on how bright the solar disk still is even with the filter. The number of sunspots that you see will vary from year to year. Actually we see a great number of spots for eleven years then hardly any spots for eleven years. As I write this book we are in a sunspot maximum and will be for several more years. If you look at the sun within this time you will see as many as fifty spots with a three-inch refractor; a four-inch refractor will show over 100 spots.

As you look at the spot you will clearly see the umbra and the lighter penumbra which surrounds it. Sunspots occur in pairs, so you will see groups of them together rather than one single spot.

Moving up to the higher powers (150X) you will be able to study individual groups of spots and you will see how they interact as material swirls around. It is a good idea to draw the location of the spots so that you can plot them from day to day as they move across the solar disk. You will be able to study and see a spot's development from day to day—from birth to death.

Another interesting feature to look for is the solar granulation that covers the entire surface of the sun. One might think that the surface would look smooth, but it does not. It gives the appearance of a coarse rice pudding. Many amateurs call this effect solar grains and is very prominent at the lower powers. It can be seen with a three-inch refractor, but a four-inch will show it much better.

To get more contrast, which will show the sunspots and granulation better, use a yellow filter or a green filter with a 25mm Kellner eyepiece. This will provide a darker background and make the features stand out clearer. If you look toward the edge of the sun's disk you will see bright streaks called faculae. They are often seen near sunspots but are normally only toward the limbs of the sun. This is because the sun, like all circular astronomical bodies, show a phenomenon called limb darkening. The limb-darkening area is quite noticeable and when the area is observed shows a great deal of granulation. The faculae show up quite well here and may even look like eruptions streaking up from the solar surface. Faculae are somehow connected with the intense magnetic fields around sunspots.

You might want to try to look at them at the higher powers since at the lower powers they might appear undefined. Use about 100X or so with the three-inch refractor or a six-inch reflector. If you are using a four inch refractor or an eight-inch reflector try about 175X. You will see how this network of bright lines and patches works its way around the spot-complex areas.

Solar flares can not be seen unless you have special equipment so I will not talk about them too much. I will say that they can be seen without the special equipment during a total eclipse of the sun. During this time only large eruptions called prominences can be seen streaking up from the eclipsed solar disk. The sun is a very interesting object to observe and the display of spots during a time of sunspot maximum is a fantastic sight. One good thing about observing the sun is that it is out there during the day and you can enjoy it in the warmth and light.

Our local group carries out a solar project during the sunspot maximum, and over a period of several years so far have acquired hundreds of observations of the formation and movement of sunspots. We also study the relation of these spots to the Earth's weather and upper-atmosphere activity. I use a three-inch refractor with the Tuthill Solar Skreen. It provides a good look at the sun. I have also taken a number of photographs of the solar disk which appear in the photograph section of this book.

Be careful when you observe the sun and follow all the safety precautions. Make sure all systems are go before you look into the eyepiece. Solar observation will be quite an experience for you if you have never seen the sun through a telescope.

Observing Mercury

Mercury is one of the most difficult planets to observe. This is because it is close to the sun and because it's just a small-sized planet.

If you want to see Mercury you will have to wait for an

elongation. That is when the planet and the sun as seen from the Earth are at right angles. At this time it will be at its highest in the sky.

Even at its highest Mercury gets no higher than twenty-eight degrees above the horizon. To get a look at this tiny planet you will have to observe from an area that has a clear view all the way down to the horizon. Have your telescope set up already and in the twilight light start looking for the planet in your finder. If you are observing an eastern elongation you will see it low in the western sky about a half hour after sunset. If you are observing a western elongation you will see it low in the eastern sky just before sunrise.

A magazine such as *Astronomy* will give information on where to look for Mercury during different times of the year.

I myself have only seen the tiny planet about five times in all the years that I have been observing. This is because my area is quite hilly and finding a location that offers a good view of the horizon is hard in this area.

You will need at least a six-inch reflector to just about see the disk—a three-inch refractor will show it but it will be hard to make out. Although Mercury is bright and visible to the naked eye as a reddish star it may be a quite disappointing object for amateurs with small and medium telescopes. An eight-inch reflector is needed to get a good look at it. You should observe with at least 200X to 300X with an eight-inch. A four-inch refractor will also show the planet well at 150X.

As you look at the planet you will see that during elongation it displays a half-moon-type phase. The planet's disk will be red or pink in color and very unsteady. It might appear at times as if you were observing it through heat waves or as if it was under water. This is because it is so low you are looking through a great deal of the Earth's atmosphere to see it. The planet suffers great instability of the image because of atmospheric refraction. This is also why it looks pink.

Most amateurs look for Mercury, observe it, then never look at it again since the disk is so small and very hard to see with a small telescope. I find the planet very interesting and worth checking out every now and then during an elongation.

Observing Venus

Venus is a grand sight even without a telescope. As a morning star it is visible in the eastern sky; as an evening star it can be seen in the western sky. Venus can get so bright at times that it can cast shadows on the ground. As seen through the telescope, Venus will undergo phases like the Moon—a very interesting sight even through a small telescope. To observe the phases, a six-inch reflector or three-inch refractor will do. Venus, like Mercury, is visible just before sunset and sunrise so the image may not be steady, due to the atmosphere of the Earth cooling or heating at this time. All you will need is about 60X to see the disk of the planet well. Actually the disk can vary a great deal in size. When Venus is approaching inferior conjunction (forming a straight line with the Earth and the sun) the disk will be quite large but the planet will show a thin crescent phase. It is also hard to observe at this time since it will be very close to the sun and very low in the sky during twilight hours.

As Venus moves into superior conjunction (the Earth and the planet forming a straight line with the sun in between) the disk will be full, but very small and hard to observe. Once again, Venus will be at its highest in the sky during elongations. The "thin" phase of Venus is a fantastic sight and can be observed and photographed even with toy telescopes.

Venus will show a yellow-white disk during elongations and you might even see some dark areas or light gray areas. We never see the surface of the planet, only its super-dense cloud cover. So, the markings that we see are nothing more than shades in the cloud cover. These shades can be seen barely with an eight-inch reflector, but you will need a ten or twelve to see them well. Many amateurs and professionals have reported these shades take on some unusual patterns—from spoke-like lines to patches with jagged shapes emerging from them. Also, during the crescent phase, many have reported seeing a darkening at the cusp of the horns. I myself have seen dark markings in the cloud cover of Venus on two occasions. The first was in 1975 during an eastern enlongation of the planet. At this time it was very high in the western sky and very easy to observe. I was using an eight-inch Newtonian reflector at 132X. This marking was a line stretching one fifth across the planet's half disk in the northern hemisphere. The second time I observed dark areas was a more recent observation using an eight-inch Schmidt-Cassegrain at 234X. The date of the observation was February 4, 1980 at 2400 hrs UT. This dark feature that I observed on the date was very similar to the observation made in 1975. As a matter of fact they both looked to be the same type of marking. I also photographed the marking and you could just barely see it in the original slide. It is not included in the photographic section of this book because the reprint will cause a great deal of image loss and the marking will not be visible. I did however include a photograph of the planet Venus during a crescent phase. This was photographed using a ten-inch reflector.

If you want to look for these dark markings on the cloud cover I suggest that you use at least 150X with an eight-inch reflector. It might help to use a yellow filter since it will provide the necessary contrast to help you see these markings. The various astronomy magazines will have dates and times of elongations and conjunctions of Venus.

In order to see the crescent phase of Venus you must observe it at least a week before inferior conjunction. You will have to go to a place that has a clear view of the horizon since the planet will be very low.

Table 1. Future Inferior Conjunctions of Venus

Year	Date	Remarks
1980	June 15	Past conjunction, very thin bright crescent phase.
1982	Jan. 21	Good time to catch phase, winter weather get's dark early. Make sure however you bundle up.
1983	Aug. 25	Start observing phase at least two weeks before date since Venus will still be in day like skies.
1985	April 3	Will be North of the Ecliptic. Before conjunction date.
1986	Nov. 5	Will be south of Ecliptic before conjunction date.
1988	June 12	No remarks
1990	Jan. 18	Venus will be brightest object in Western sky.
1991	Aug. 22	Look also for Venus at sunrise close to the Sun.

Observing the Moon

The moon is the closest planetary body to us in space. It is the long-time favorite of amateurs who want to impress their friends with the beauty of the universe. Even a pair of binoculars will show the larger craters and the so-called seas which appear as dark patches.

Through a telescope as small as two inches a great wealth of detail can be seen, but for a magnificent view use a six-inch reflector or a three-inch refractor. The larger the telescope you use the more fantastic a view you will get of the vast network of lunar mountains and valleys and craters. An eight-inch reflector provides a view you will never forget. As a matter of fact, no matter how many times I have seen the moon through telescopes of all sizes I never lose interest in it. Start out at low-power—about 50X—when the moon is in a phase close to first quarter. Shadows thrown by the sun on the lunar formations will give depth to your view. Also at this time the moon is not very bright and the detail will not be washed out. Contrary to what many people believe the full moon is not the best time to observe the moon. There is a saying among amateurs and professional astronomers alike: "When the moon is full we take the night off." This is true since the full moon is so bright it lights up the whole sky. All other objects are subdued by the lunar light including the majority of lunar formations. At 50X the entire moon should fit into the field of view and you will see a breathtaking amount of detail.

The craters of the moon are the product of meteor impacts which occurred quite a long time ago. Some of the craters are also extinct volcanos and these can be seen as having dark basins in the center of them. One classic example is the crater Plato which is located in the northern hemisphere of the moon. This crater is of volcanic origin and is quite large, being over sixty miles across.

Move next to the higher powers (150X to 500X). At this power you can examine each structure separately.

Since there is so much to see on the moon an observer may and can spend an entire night observing just the moon. It would be impossible to cover all the major details of what you can see since the length of this book does not permit it. There are a number of books out that cover only amateur observations and photography of the moon. I will however go over some of the most interesting prominent lunar formations.

The Crescent Phase

At this time you will be able to observe a very interesting phenomena known as Earthshine. This is the moon being illuminated on its unlit sky by the Earth's reflected light. Through an eight-inch or ten-inch reflector you may even see detail of lunar formations along the unlit side. When this phenomena is observed at the crescent after last quarter it is called "the old moon in the new moon's arms."

Observing the crescent phase at about 100X you will see a large circular basin or sea called the Mare Crisium. The area in this "sea" is very dark; the basin is over 300 miles in length. It will look good at about 200X since around the rim is a very mountainous area.

South of Crisium will be a larger but flatter dark basin called the Mare Foecunditatis, a very large sea separated from Crisium by a high rough area. Around this area are also a great multitude of old worn-down craters. One of the larger ones is called Vandelinus and is 100 miles in diameter. It can be seen southeast of Foecunditatis at low powers but use 100X or more to get a good look at it.

Remember if you are using a reflector to observe the moon, south will be up. East will be to the left, north will be down and west to the right. If you are using a compound telescope or a refractor with a star diagonal then north will be up and south will be down.

Looking north of Crisium we see the crater Macrobius. This is about forty-five miles across. Most of the area just north of Crisium is, once again, speckled with small craters and hills. Most are quite worn down and flat.

If we follow the curve of the crescent further north we will come to two very interesting craters, Hercules and Atlas. Hercules is more towards the east and if you look at it under a higher power you will see that it is a high-walled structure. Also, you will note that inside it is another crater. Close by, toward the west, is the crater Atlas. Atlas is a little larger and is fifty-five miles in diameter.

First Quarter

By waiting a few days more a great amount of objects have now become visible as the disk of the moon increases in size. This is one of the best times to observe the moon since the sun will still be low in the lunar sky. This will provide a great deal of shadowing effects which will make the formations stand out very well.

You may start with 50X or so to get a view of what to expect. At this power you will see everything in a sort of wide angle view. You may see something that might spark your interest and then you will go the the higher powers to take a close up look at it.

South of the moon you will see a great amount of rugged terrain. This is part of the southern lunar highlands that you are observing. This area has so many craters that they overlap each other.

Just north, we see a small somewhat circular dark basin, this is the Mare Nectaris and in the approximate middle is a high-rim crater by the name of Fosse. Just off to the east we see three large craters forming something of an arc. These are very close together. The larger one which is close to the bay area of Nectaris is called Theophilus and is sixty-five miles in diameter. If you look closely you will see that Theophilus has a very unusual formation in the center of it. It will look like a small hill or mountain. Quite a few lunar craters have a similar structure and astronomers think that they are caused by a piling up of lunar material when the crater was made by a meteoric impact.

Going north of the Mare Nectaris we come to the start of the Mare Tranquilitatis or the Sea of Tranquility. This is perhaps the most famous sea on the moon since it is here that the astronauts of Apollo 11 landed and Neil Armstrong made his historic "small step . . . ."

As observed from Earth with a moderate-size telescope Tranquility is quite smooth. A larger telescope will reveal a number of small craters and small mountain ranges, especially near the edges.

Tranquility is separated from Nectaris by a group of highlands. Here we also see the crater Madler which has a diameter of twenty miles. Although it is not a large crater it is quite prominent since it is in a section of area which has a high elevation in comparison to the areas around it. Going northward Tranquility blends into another large lunar basin called Mare Serenitatis or the Sea of Serenity. Mare Serenitatis is almost the same size as Tranquility and is very smooth. There are a number of bright rays which stretch through it that lead to the southern lunar highlands.

Going north we come to the end of this sea. You will see some very high areas and they will look quite brilliant as compared to the low level basins. The area that marks the end of Mare Serenitatis is the Caucasus Mountains area and is a great sight even at low powers. You will see in this area many mountains of many different sizes stretching for hundreds of miles across the lunar surface. The average height of this mountain area is 12,000 feet.

If you look just north of these mountains you will see a truly remarkable crater. This is Aristoteles and is sixty-five miles across and 11,000 feet deep. When the sun is striking the rim of this crater the shadow of the rim can be seen inside the crater blocking half of it out.

Just north of this crater we come to another sea. This is the Mare Frigoris and is the longest sea on the moon. It is very narrow, but runs across the whole northeastern section of the moon. At this phase you will only see part of this sea.

The Gibbous Waxing Moon

This is the moon about three-quarters full. At this time it will be quite bright and only objects several hundred miles west of the terminator will show up well.

A low-power view of this phase is well worth the time. You will see many high features reflecting the sun's light in the unlit side close to the terminator (also called the twilight zone, not to be confused with the TV series where strange things occur). This is the line that separates day from night. The best amount of detail can be seen along the terminator at this time since this is where the greatest shadowing effect occurs.

You will notice even at the lower powers the difference between the southern part of the moon and the northern part. The southern part of the moon is very rugged and cratered. It is often referred to as the southern lunar highlands. Once again south will be up if you are not using a star diagonal; now switch to a higher power like 200X or so to get a close-up view of some of the largest craters on the moon.

Starting from the southern lunar highlands, about as far south as you can see will be a very large crater. If you look closely this crater will have several smaller ones in it that form sort of a curve towards the east. There should be three of them, also quite large, all in a row. This large crater is called Clavius and is 145 miles across. If you were standing in the middle of this crater you would not see the rim since it would be below the lunar horizon. You (if you didn't know it to begin with) wouldn't even know that you were in a crater. Clavius is the largest crater on the moon and is easily visible in the smallest 'scopes. Just north of Clavius is another large crater, also a great sight in even small telescopes. This crater is called Maginus and is 110 miles in diameter. Just east of Maginus is the large crater Longomotanus which is ninety miles across. There are quite a few large lunar craters here, but the most prominent ones I have mentioned above.

Just north of Maginus however is one of the most spectacular lunar craters, Tycho. Although it has a diameter of less than sixty miles it is a very high rimmed crater and very young (in comparison to other lunar craters). Tycho has a great many bright rays emitting from it in all directions caused by material from the lunar surface being thrown out in all directions during the meteoric impact that created the crater. Tycho can be seen with binoculars and also can be seen with the naked eye. A low-power view will show how the rays streak out in all directions across the moon. If you look closely you will see that they reach out for a great distance hundreds and hundreds of miles from the crater.

Directly north of Tycho you will see bright lines in a large basin or sea. This is called the "straight wall" and is the most interesting sought after feature on the moon. It is a mountainous wall 800 feet high. At low powers it may look like a bright line, but at the higher powers, when the sun's angle is just right, it will be a sight you will not soon forget. It's a feature you will show your friends during a star party. The sea that the wall sits in is the Mare Nubium and in this sea some of Tycho's rays can be seen streaking through it. Scanning further north you will come to the first of three large craters in a row. They are a great sight in a six-inch at about 150X. At this power you should be able to fit in all three which should look fantastic if the terminator is close by on the date of your observation. You will see that they curve somewhat toward the west and they are on the shore line of the Mare Nubium. The furthest one south is a high-walled crater by the name of Arzachel. This is sixty miles across and over 13,000 feet deep from the top of the crater rim. The next in line is Alphonsus which is a larger crater about seventy miles across, but only about 5,000 feet from the top of the rim.

The third in line, Ptolemaus, is recognized easily since the floor of this crater is almost as dark as the Mare. Ptolemaus is one of the most prominent of all lunar craters because of its size and the dark interior. Ptolemaus is about ninety miles across and 9,000 feet deep. It is most likely of volcanic origin. Use as much as 400X with a six-inch or 500X with an eight-inch 'scope to study the internal structure of the crater and the high mountain like walls around it.

Just below (if your image is inverted) you will see another crater and although it is small in comparison to the others it stands out very well. This small cone-shaped crater is called Herschel and is twenty-eight miles in diameter. Moving further north you will be seeing quite a number of small mountains, valleys, and craters of all sizes. Then you will come to a smooth circular area. This will look somewhat strange since it will be surrounded by high areas and mountains on all sides. This is the Mare Vaporum or the Sea of Vapors. Its floor is dark in comparision to the bright lunar high grounds like the rest of the Mare, but it is very smooth. If you study the floor of this Mare you will see a number of rolling hills. You'll need at least an eight inch reflector to see them at about 200X or more.

Just north of the Sea of Vapors you will see a very large

mountain chain. These are called the Apennines and are almost 20,000 feet high. They are also referred to as the Himalayas of the moon. They are so large that they look three-dimensional in contrast to the lunar surface. One person who was looking through my telescope at them remarked, "This is fantastic. It seems as if you can reach out and touch them." You will see how fantastic they are. I suggest that you observe them for best effects at about 100X. You will notice that the shadows they throw on lunar surface are quite long and dark. This is a tribute to their height.

If you follow the mountain chain to the east (or right) you will see them lead right up to a very high crater. If you are using low power at about 50X it will seem as if the mountains are going right in the side of this crater. As you use more magnification however, you will see that there is a good distance of separation to where the mountains meet the crater. This crater is called Eratosthenes and is forty miles in diameter and being 16,000 feet high it blends in perfect with the mountains close by. Take a close look at Eratosthenes at about 300X. You will notice a mountainlike structure in the center of it. Also around the crater rims are small mountain chains. Some faint and bright rays can be seen also stretching out a small distance from the crater from all sides. It might be hard to see them since nearby Copernicus with its great bright ray system runs through the entire area.

Copernicus is just southeast and is a fixty-six-mile-wide crater from rim to rim. It is 17,000 feet high from the base (or deep, from the top rim, whichever way you want to look at it). The thing that makes this crater stand out is its very bright ray system. The bright rays coming from this crater can be traced all over the east-central part of the moon.

At the northern base of Copernicus begins the Mare Imbrium, the largest sea on the moon. It is somewhat circular in shape and if you look closely you will be able to see the rays from the crater Copernicus streaking through it. Imbrium is quite dark, but the surface is covered with a great many small creaters which can be seen at about 200X or higher.

Boarding on the western shore of the Imbrium basin is the crater Archimedes which is fifty miles across. This crater has a dark floor to it which makes it stand out very well. Although the floor is not as dark as the floor in the crater Ptolemaeus it nevertheless is an interesting object.

Going further north the number of craters decreases and we begin to see more smooth areas with some high mountains. If you scan further north you will see what appears as if someone has just taken a knife and made a deep cut into the lunar surface. The Alpine Valley is about eighty miles long and eight miles wide. This valley was created from a glancing meteoric impact.

The Alpine Valley lies almost in the lunar alps. You will observe that the valley is surrounded especially to the southeast by very rugged terrain. The Lunar Alps are another fantastic mountain chain. Although they aren't as high as the Apennines they are almost 12,000 feet high.

All this time you probably have been seeing a large crater with a very dark floor in the north. You most likely saw it when you observed the Alpine Valley. This is the volcanic crater Plato. It lies northeast of the Alpine Valley and is sixty miles across. What makes it so visible is the fact that it lies on bright lunar grounds 7,000 feet high. Surrounded by Imbrium to the south and the Mare Frigoris to the north, this small piece of high ground stands out. Plato in the middle with its dark floor adds greatly to the scenery. Plato has a jagged rim which can be observed at 300–400X with an eight-inch or ten-inch reflector. It can also be seen with a four-inch refractor with difficulty at 200X. Just southeast of Plato in the Mare Imbrium are the Teneriffe Mountains. This small mountain chain is seen as bright specks on the dark Imbrium floor at low powers. With a six-inch reflector at 200–300X they start to appear like mountains and you may see some hint of the elevation, but it will take at least an eight at 400X to see them well. These mountains are just under 8,000 feet high for the highest ones. At times if the sun is falling right on them they take on the shape of a ship's anchor.

Directly east of these mountains is a bright jagged structure, also on the Imbrium floor, called the "straight range." This is a rugged, mountainous region 6,000 feet high. Also around this region of the Teneriffe Mountains and the straight range are a number of peaks. These will appear as bright dots on the dark Imbrium surface in a small telescope; however a larger scope of ten inches or more will start to show some elevation. These peaks are 5,000 to 7,000 feet high.

Observing the Moon Several Days before Full

Most of the features at this time will be washed out from the intense lunar light and you may want to wait until after full Moon when it starts to wane (or starts to get dark from east to west heading towards last quarter) before you observe it.

In the south, all of the Mare Nubium will be seen. To the far east will be a circular small mare called the Mare Humorum. Just north of it to the far east is the Oceanum Procellarum.

Just about in the Oceanum Procellarum is the bright rayed crater called Kepler. Kepler is located just east of Copernicus and although it is only twenty-two miles in diameter it is very high. Just north of here is another bright rayed crater which will catch your eye. This is the crater Aristoteles and is sixty miles across and almost 12,000 feet deep. Both craters mentioned above stand out very well since they are of the bright rayed type and they are sitting in or close to the dark Procellarum basin.

If you look further you will see a great many other craters of all sizes. You will notice mountains, valleys, and so on. At this point however, since the sun is almost directly overhead on the moon, there will be little or no shadow cast, and the depth and dimensional effect will be lost. The high areas will look bright and the lower areas will look darker.

The Full Moon

Use low powers only since what you will see is for appreciation of the beauty of the moon and not for serious lunar study. Many amateurs photograph the full moon and make an enlargement of it to hang in their observatories or homes. The full moon is very easy to photograph since it is large in apparent size and is very bright. Because it is bright you might need some filteration such as a dark yellow or green filter. This will however give you a false color. There are several types of lunar filters which can be used successfully on the moon. Some of the smaller refractor such as the 60mm or the three-inch will come with a special filter for observing the moon. These filters cut down some of the light so the very bright image doesn't bother your eyes. The moon is so bright with an eight-inch reflector that you can project its image on a white wall or screen five feet away.

To observe the full moon a small refractor or moderate refractor will work better than the reflectors. This is because the reflectors gather too much light and with the reflecting of the image to the eyepiece there is often considerable glare.

The Waning Moon

Compare the features some time when the sun is throwing light at different angles on the lunar landscape. Compare the appearance of the lunar features from first quarter to full, then from full to last quarter. You will see that the angle of the sun in the lunar sky makes quite a difference in how the lunar landscape appears. Some features even appear to change shape as seen from the Earth. Strange reflections and casting of shadows produce unusual-looking formations.

There was a book published a number of years ago called *Somebody Else Is on the Moon*. It was written by a very imaginative fellow by the name of George Leonard. Mr. Leonard claimed that an extraterrestrial race from another star system is using the moon for a base and for mining operations. He illustrated his book with "official NASA photographs." Mr. Leonard points to a number of structures and shadows and bright areas on the Moon and claims they are a superrace's machinery. I have heard the story of so-called lunar bridges, domes, and so on. I have often photographed the phenomena that Mr. Leonard says are of alien construction. I have done a study of the moon visually and photographically for a number of years. I am convinced that all these structures are natural casting shadows and reflecting the sun's light. As I have stated before, the shadowing and reflection effects from the sun on the lunar features can look highly unusual and you can imagine them to be bridges, domes, or whatever you want.

For example, you may see a strange-looking shadow on the moon one night, and the next night it will have changed shape or even become invisible. This is because in that one day's time the sun has moved from its previous posistion in the lunar sky. This will cause an altering of the appearance of the lunar surface.

The moon is not a dead world, geologically. Our astronauts have recorded "moon quakes." Also, throughout time, smoke and lights or flashes have been seen in the volcanic craters of the moon. Mr. Leonard might point out that this is some alien machine. I feel that this could be evidence of volcanism on the moon. Most of these transient features like the flashing of light, pink or reddish smoke, bright areas, and so on have been recorded by amateur astronomers.

I will never forget the time I was observing a third-quarter moon in June of 1977. I was using a ten-inch reflector in my observatory at about 80X. As I was looking at the moon a small silver spherical object, shiny but somewhat transparent, caught my eye. It started on the eastern limb of the moon and was slowly moving towards the western limb. It was moving in an erratic fashion right across the center of the disk. The moon was very bright and it was somewhat hard to follow it. I watched it knowing that it would soon move from in front of the moon and this would allow me to get a better look at it. My excitement grew. *After all my years of looking I'm finally going to see a UFO.*

As it approached the end of the moon my heart started beating wildly. *Any second now*, I thought to myself. Then it came out from in front of the moon and I got a good look at it. My excitement ended; it was only a weather balloon. It did look strange crossing in front of the moon like that, but the best part of a mystery is solving it.

Observing Mars

To the naked eye Mars looks like a rusty-red colored star and during times of opposition it can be very bright. Mars comes into opposition every two years, but most closely once every fifteen years. A favorable opposition means that Mars and Earth will be very close at this time. An opposition of this kind occurs during July and August only. This is because at this time it is summer on the Earth for the northern hemisphere, thus the Earth is at its farthest point from the sun. On Mars at this time it is winter in the northern hemisphere and at this time Mars is at its closest to the sun.

Before we get into observing Mars I would like to give some information about the planet since it is a most remarkable object for the amateur to study. Until the day of the space probes Mariner and Viking, little was known about Mars. In the 1930s it was thought to be the home of a race of beings who in a last desperate attempt to save their race were draining water from the polar caps by digging the Martian "canals." Many professional and amateur astronomers have reported seeing the canals, but the observation of this phenomena was limited to only times of excellent seeing conditions. Even when conditions were perfect they would only fade in for a second or so then fade out once again, perhaps for hours. Because of this, many began to believe that they did not exist at all and that they were nothing else more than an illusion caused by the distortions in our atmosphere and overactive imaginations.

We know that the canals do not exist, but something else does. The space probe Mariner 9 which was launched on May 30, 1971 gave us our first really good look at the Martian surface. Mariner 9 reached the red planet and orbited in on November 13. The probe took countless television photographs of a great network of meandering channels which are now thought to be extinct rivers. Could these river channels be the Martian canals that were reported from Earth? It is possible, but no one can say for sure. I heard a story sometime ago which can perhaps shed some light to the above question.

An amateur astronomer who owned a twelve-inch reflector was observing Mars during a favorable opposition. As he observed the planet he thought he saw two large craters. Later the next day he went to visit an astronomer friend of his at a well-known observatory and they started talking about Mars. The amateur said, "You know last night I was looking at Mars through my telescope at 400 power and I thought I saw two large craters. One was larger than the other, but they faded in and out and I could not verify it."

The astronomer who was quite famous at the time looked at him with excitement and said, "Yes I have seen it also on several occasions. It would be visible for only a very short time and then fade out. I didn't want to say anything to my coworkers because I thought they would think I was seeing things."

The professional astronomer was the great E. E. Barnard. The date of the conversation was sometime in 1891 at the Lick Observatory in California. The name of the amateur astronomer was never recorded. The information was taken from Dr. Barnard's diary.

The Mariner probe sent back pictures of a great volcanic crater now called Olympus Mons (Mount Olympus). This enormous structure is 375 miles across and sixteen *miles* high (84,000 feet high from the base). Could this be the crater seen by Dr. Barnard and his astronomer friend? Today most scientists think so. Olympus Mons can be seen from the Earth quite well in a telescope of ten inches or larger. It can also be seen in smaller 'scopes such as an eight-inch reflector or a four-inch refractor, but it would look only like a bright yellow patch. Near Olympus Mons are two more large volcanic craters. The first is called Arsia Mons, the other is called Pavonis Mons. They are also large but no where as large as Olympus. Could one of these be the other crater mentioned in the story?

Yes, these areas can be seen without too much trouble with large amateur telescopes during a Martian opposition. There are two reasons why we see them more clearly today. The first is because we know that they are there and we know what to look for. The second reason is that because the optical quality of telescopes today is much better than it was seventy years ago or so, we get much better resolution.

Mars is a geological wonder. The Nirgal Vallis is a meandering channel 600 miles long, and as seen from the Earth it could look like Martian Canal. It was once thought that the Martian polar caps were composed of solid carbon dioxide, but recent research suggest that this is not completely true. It is now thought that only the outer fringes of the cap are composed of dry ice while most of the cap is made of water, in the form of a permafrost similar to that which is found on the Earth.

The Martian atmosphere is composed of mostly carbon dioxide with some water vapor and nitrogen. The soil is composed of mostly liminite with great amounts of oxygen and water trapped in it. The combination of this gives the Martian surface its reddish appearance.

Another interesting feature on Mars which can be seen from the Earth is the Valles Mariners. It is 150 miles wide and four miles deep. Valles Mariners is called the Grand Canyon of Mars. It is however a great deal bigger than the Grand Canyon of Earth which is "only" eighteen miles wide and one mile deep. It is located near the Martian equator and will appear as a darkish line irregular in shape.

Is there life on Mars? This was to be answered by two American space probes sent to Mars. Viking I landed on July 20, 1976 in the Chryse Planitia area of Mars. Viking II was launched on September 9, 1975 and landed in September of 1976 in the area of the Utopia Planitia. The purpose of launching two probes is that since this is a very important project if one probe failed we would still have another chance with probe two.

In both Vikings there were two basic parts: the orbiters which were to circle Mars and send high resolution photographs of the Martian surface back to Earth. The second parts were the landers which were to land of the surface and perform a number of experiments.

The Landers

The surface of Mars is quite cold. Temperatures range from minus 122 to minus 22 degrees Fahrenheit. It was found that the atmosphere has a pressure of only five millibars. With these conditions it seemed very unlikely that life could exist on the surface of Mars. The most exciting part of the mission was the search for life in the Martian soil. At first the results were highly promising but now they are very uncertain. No one will say for sure if the results of the biology experiments were positive or negative.

The biology experiment on the lander was a three-part one. One part took a sample of the Martian soil, warmed it, then added a nutrient that was slightly radioactive. If the microorganisms did grow from the Martian soil they would then consume the nutrient and give off radioactive waste gases. These gases would then rise above the chamber and be detected by an instrument.

The second experiment measured the amount of oxygen that formed in the tube to detect microbes that feed on carbon dioxide and give off oxygen as a waste product.

The third was the detection of organic molecules in the soil using a special device called a gas chromatograph mass spectrometer which could "sniff out" the presence of dead microbes. Although the instrument showed no organic molecules present in the Martian soil the Viking scientists were not dismayed, since the sensitivity of the device was in question. However the data from the first two experiments caused an initial flurry of excitement. Each indicated that something was taking place inside the chambers. The chamber that measured the formation of oxygen reported a fifteen-fold increase in only two hours. The results looked promising and even the test for the radioactive waste gases proved positive. At this time an army of reporters was ready to break down the door of the control room of the Viking scientists. They had to make some type of statement to the press, so a meeting was held. At the meeting all the evidence gathered thus far was weighed and many of the bolder scientists were ready and willing to make a statement to the press and the world that life existed on the surface of Mars, in at least microbe form. Advisors and some of the senior project scientists thought it might be a good idea to hold off in case that they might be wrong. Another factor was how would the public handle it? It was safer at the time and easier for them to simply state that the results of the experiments were inconclusive. That's exactly what they did and till this day NASA states that the "biology experiment of Viking I and II are inconclusive."

It might be interesting to note that before the results of the experiments came in all the scientists agreed that if such results were obtained that it would be proof enough to them that some type of life did exist in the Martian soil. The results did prove positive beyond their expectations, yet as of 1980 no more is heard of the Viking biology experiments. Many believe that the probes did find life, but that NASA and the government are withholding the information until they can determine what effect it will have on the population of the world.

Viking was shut down during the Martian conjunction of 1976 then turned on once again in November 1977. An attempt to redo the biology experiments failed because the intense cold of Mars at the time, –200°F, did damage to the biology chamber of both landers. At least this is what they claim.

Viking is now turned off and remains dead on the Martian surface where it will stay perhaps forever. The orbiters are still in operation and recently photographed an intense storm on the surface. The type of storm seen is very similar to a hurricane on Earth, but much more intense due to the thin Martian atmosphere's great wind velocities.

Future trips to Mars include two "rovers" which will be run by solar energy by remote control from the Earth. This rover will have a camera so the scientists on Earth can direct it to places that look interesting.

Observing Mars through amateur-sized telescopes can present a problem. You can see a great amount of detail in telescopes ten inches and larger, but even so, observing the red planet is not an easy task. One of the reasons is that our atmosphere will cause the image of Mars to be unsteady and blurred at times when there is even a slight wind. This wind may occur in the upper atmosphere. The air currents of our planet cause an instability in the atmosphere, thus causing an unstable image of Mars especially when you are using the higher powers. When the atmosphere reaches a moment of stability when there is no wind then the surface details and other features of Mars will come in quite sharp and steady. This image stability many last anywhere from three seconds to several minutes until the planet's detail is once again distorted by the atmosphere.

The amount of time that the image will stay steady and clear will depend strongly on the state of the atmosphere at the time. A very poor time to observe the planet will be when a storm is less than twenty hours from your location. Also twenty-four hours after a cold front has passed the air will be very unstable although the skies may be very clear. This is because a cold front brings with it northwestern winds (if you live in the New England area). These northwest winds are at low and high altitudes and will cause a great atmospheric instability.

The best time to observe the planet is when it is at opposition during the summer months. During the summer the air is quite stable and you usually have a slight haze. It is best to look at Mars on a warm summer hazy-clear night. These are ideal conditions since the haze will mean air stagnation and no wind.

If you are planning to look at any planet but especially Mars, always wait until the planet is at least forty degrees high in the sky. Looking at it low in the sky will also cause distortion to the image since you will be looking through a greater amount of atmosphere. To get the best view wait until the planet reaches the meridian or its highest point in the sky. Here you will be looking through less atmosphere and you will be rewarded with the best possible observing for the night.

Mars displays a very small disk and the only time that you can observe it without trouble is during opposition. As I have mentioned earlier, favorable oppositions occur every fifteen years, with the next ones occurring in 1988 and 2003. But from 1984–1988 the planet will also be in a good position for observing and will show a fair-sized disk. Since Mars comes into opposition with the Earth every two years the planet will be close to us in 1984, 1986, and finally (closest) in 1988. The apparent disk size of Mars as seen through the telescope will vary a great deal depending on the oppositions. It is during these favorable opposition that all amateur telescopes are pointed at Mars.

A six-inch reflector or a three-inch refractor will show only the major features during the opposition. One of the features which is quite visible is called the Syrtis Major. This is a large teardrop-like feature on the surface and is very dark. Many have reported it to be a black or a dark green. It extends northward off a large long faint dark area which you may see only during fleeting moments of good seeing. With the smaller telescopes you will also see an arm-like dark green feature extending westward from the Syrtis Major region. This is called the Sabaeus Sinus. It has an unusual claw-like extension which points northward on its western end. You may see it as an oval patch of dark areas with the six-inch.

Moving toward the south the large basin Hellas will come into view. Hellas will appear as a yellow-white circularly shaped area. It is located directly south of the Syrtis Major.

The northern (or southern) polar cap will also catch your eye. Through a six-inch reflector it will look like a gleaming white spot on the pole of Mars. At times the cap may be so bright that it may look like it's floating above the pole. But what cap will be visible at what opposition? If opposition occurs during the winter for the northern hemisphere on Earth then it will be summer on the northern hemisphere of Mars. This will be an unfavorable opposition since at this time Mars will be farthest in its orbit from the sun.

During this time the northern cap will be visible but it will be very small. The southern cap will be quite large but you may have difficulty seeing it. When opposition occurs when the northern hemisphere of Earth is in summer then the northern hemisphere of Mars will be in winter. This is a favorable oppostion since at this time the Earth will be at its farthest approach from the Sun and Mars will then be at its closest approach to the Sun. At this time the northern cap will be large, but the southern should also be visible.

Using a moderate-sized telescope—a reflector of eight inches or a refractor of four or five inches—you get a clearer and better look at the features above. You will see without too much trouble the claw-like extension off the Sabaeus Sinus. This is called the Meridiai Sinus and is composed of two finger-like projections facing northward. Also you may begin to see more bright areas. One area will be just west of the Syrtis Major—called Aeria it is just barely visible through an eight-inch reflector.

Above the Syrtis Major you will see long dark patches extending in all directions. Although these patches are all connected as one solid body you may not see the connection points through a five-inch refractor or an eight-inch reflector. One of these areas directly above the Syrtis Major is called the Mare Tyrrhenum. It is a greenish or sometimes brownish area that will be quite easily seen through an eight-inch reflector. West of the Sabaeus Sinus will be a triangular dark area called the Mare Erythreaum. It will be somewhat faint in an eight or a five and will only be visible during moments of complete atmospheric stability. Below this is a patch called the Margaritifer Sinus. Although this feature is separated from the Mare Erythraeum you may see it as one fused area since an eight-inch reflector or a five-inch refractor will not resolve them so they can be seen as two different areas.

Moving to a ten-inch or a twelve all the features mentioned above will be very prominent and clear (during times of good seeing, of course). With this size telescope you move into a different class and a serious professional study can be done on the planets. Features may be mapped and changes in these features will be obvious. Atmospheric ice clouds and dust clouds will become plainly visible. Also such structures as the Great Olympus Mons will be seen, as will several of the other large volcanos that surround it. The melting lines can clearly be seen radiating from the polar cap. The melting lines are dark rings that appear around the area of the cap that has melted. They can also be seen with a four-inch refractor or an eight-inch reflector, but with great difficulty.

Every Martian opposition *Astronomy* magazine publishes a map of Mars for the amateur observer. They also list times and dates noted features will be visible from the Earth. The address of this fine magazine can be found at the end section of this book.

It is clear from what I am saying above that in order to fully appreciate the planet Mars and do serious work you will need a reflector of at least ten inches or a refractor of five or six inches. During a favorable opposition an eight-inch reflector or a four-inch refractor will perform very well. These oppositions occur rarely, so most of the time a large amateur telescope will be needed. A ten-inch reflector is more or less on the borderline of being classified as either a large 'scope or a medium sized 'scope, so you might see that I refer to it in both of these categories.

Use an orange or light red filter for observing surface detail, but use no filters to look at the polar caps. Sometimes no filter works for picking out dust storms and clouds since they may at times provide too much contrast.

During the opposition of 1973 I was using an eight-inch reflector on Mars with about 300X or so. The time of the observation was about 5:20 UT on the sixteenth of October. I was looking at the area of the Margaritifer Sinus, Sabaeus Sinus, when I noticed an unusual yellowish cloud that was partially obscuring the Margaritifer Sinus. It was about midnight EST and I had been up late several nights before during this opposition to observe Mars so I thought to myself, *I must be tired and seeing things.* I took about half-hour break since trying to see what was going on at the time on the planet was putting a strain on my eyes. I came back outside to the telescope at about 6:00 UT and once again pointed the eight-inch Newtonian reflector toward the red planet. I slowly focused; I was shocked! A great part of the entire surface of the planet was covered by a thick yellowish cloud. Only some areas showed through, but they too were very hard to see. Almost the entire Margaritifer Sinus and sections of the Sabaeus Sinus and Sinus Meridiani were obscured.

What I observed was a great dust storm on Mars and I recorded it in my notes as such. This storm was visible on the planet into December. I had seen the storm at least twenty minutes before it was officially recorded by the twenty-four-inch reflector at the New Mexico State University Observatory.

I will never forget the sight of this ominous dark yellow dust cloud sweeping across the surface of the planet. For a while the red planet (in some areas) looked like the *yellow* planet. Storms like the one described above are often seen first by amateur astronomers during a favorable opposition. The opposition of 1973 was almost as good as the opposition of 1988 will be.

During the 1973 opposition a number of new features were seen by professional and amateur astronomers across the world. These new features were added extensions of known dark areas that were visible for as long as several weeks. They however slowly vanished after this and till this day no one knows for sure of how they were created. I myself recorded a number of new known features during the great opposition of 1973.

On October 8, 1973 at 4:00 UT I observed a most unusual feature. I was using an eight-inch Newtonian reflector at about 400X with an orange filter. What I saw was a very dark line stretching from east to west across three-quarters of the planet. As I observed this I was puzzled. I thought to myself, *Hey wait a minute. There is no such feature on the Martian surface.*

As I observed it further I saw that it ended in a dark green irregular patch near the western horizon. Cloudy weather halted all observations for the next few days, but when it finally cleared up and I once again turned the telescope toward Mars I could not see the new marking. I thought at the time that it might be on the other side of the planet. The new marking was not to be found and I could not find it the rest of the 1973 opposition. I had asked around and wrote a few places to find out how many people saw it, but few did. Nothing was written about it in any of the astronomical magazines or journals.

I came across a passage in an outstanding book on Mars which talked about the appearance of a new feature during the 1973 opposition. This feature was a dark linear line, very black in color and extended over 3500 miles across the Martian equator. This was without doubt the new feature that I saw on the date of the observation. One of the reasons that I could not see the new feature after that was the fact that the great Martian dust storm was forming. This storm was so intense as I mentioned earlier that it could have obscured the new marking. It could have also covered parts of it, altering its shape as seen from the Earth. I did at this time record a number of very dark broken patches which I feel could have been this new feature.

When the great storm finally cleared up several months later, Mars was well out of opposition and the distance between Mars and the Earth was increasing every day. The disk as seen from the Earth was getting smaller and smaller and the surface detail more vague. I never saw this marking again, but I hear that it was still seen for several months by the larger professional telescopes at a number of universities across the country. The marking then slowly faded as Mars reached conjunction and was never seen again.

Table 2. Future Martian Oppositions

Year	Month	Day	Remarks
1980	Feb.	26	Unfavorable opposition
1982	March	31	Unfavorable opposition
1984	May	11	Fair opposition
1986	July	10	Very good opposition
1988	Sept.	29	Favorable opposition
1990	Nov.	24	Good opposition
1992	Dec.	31	Fair opposition
1995	Feb.	12	Unfavorable opposition

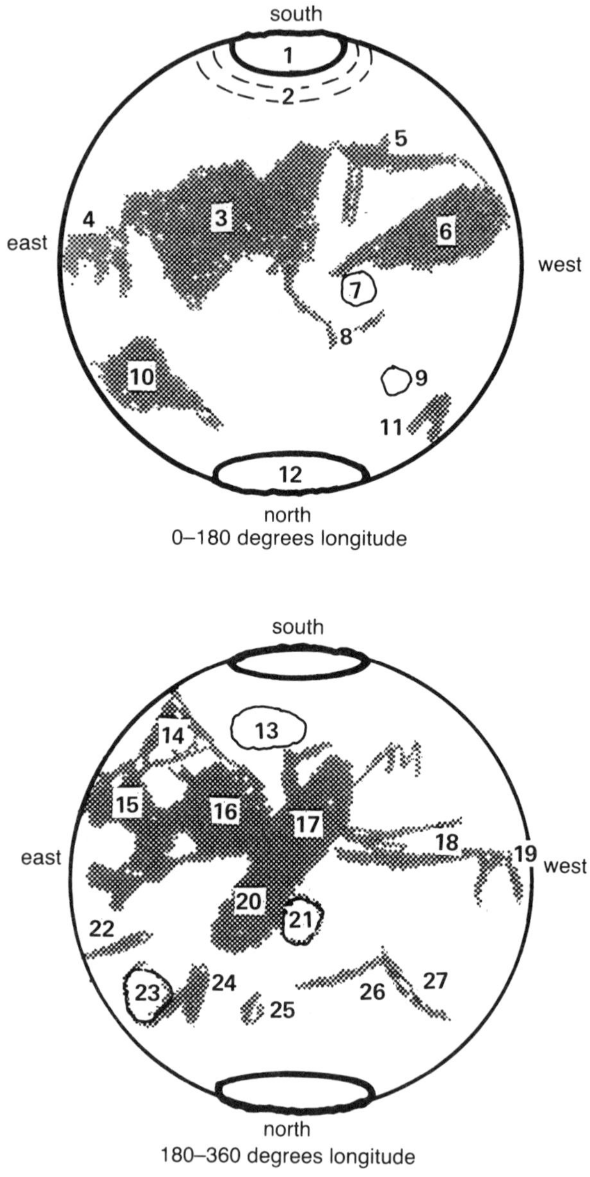

Key
1) South polar cap
2) Melting lines
3) Mare Erythraeum, Margaritifer Sinus
4) Meridiai Sinus
5) Mare Australe
6) Mare Sirenum
7) Memnonia
8) Phoenicis Lacus
9) Olympus Mons
10) Mare Acidalium
11) Euxinus Lacus
12) Northern polar cap
13) Hellas
14) Mare Hadriacum, Eridar
15) Gomer Sinus, Tritonis Sinus section of Mare Sirenum (far east)
16) Mare Tyrrhenum
17) Lapygia, Deltoton, Sinus, Mare Serpentis, Yaonis Regio
18) Sabaeus Sinus
19) Meridiai Sinus (Section of, far west)
20) Syrtis Major
21) Aeria
22) Trivium Charontis
23) Elysium
24) Aetheria
25) Coloe Palus
26) Ismenius Lacus
27) Cydonia

Figure 5. The Planet Mars. Location of surface features on the planet Mars as seen with an eight- or ten-inch reflector during a favorable opposition.

Observing Asteroids

Finding an asteroid is not an easy task. You will have to be able to use your setting circles with great efficiency to locate them. Some of the brighter ones are a little easier to find, but when you do find them all they will look like is a star.

To observe the asteroids you must use no more than 100X and the actual observation is not of the asteroid itself but of its movment. Once you do find one you must observe it from night to night and make an accurate record of your observation. You will notice than as time goes on from one night to the next that the star-like object that is an asteroid is changing its position in relation to the background stars. This apparent faster motion is due to the fact the asteroid is closer to the Earth than the background stars.

For visual observation an eight-inch reflector will do very well, however a six-inch reflector should also be adequate. If you plan to photograph then I suggest that a four-inch refractor be used since with this instrument the stars will not have spiked images and the glare of bright nearby stars will be reduced.

Asteroids, like the planets, orbit the sun so at times they will be much closer to the Earth and easier to observe than at other times. Magazines like *Astronomy*, for example, carry excellent charts to locate one of these objects when they are making their closest approach to the Earth.

Here is some basic information on the asteroids. Although there are thousands of them between the orbits of Mars and Jupiter, information on the five largest is given here. How bright an asteroid will get depends on its size, reflectability, and distance from the Earth. The scale that is given in table 3 is the magnitude scale. A star with a magnitude of six will just be visible to the naked eye on a clear dark night away from all street lights. With an eight-inch telescope you can see a star as faint as magnitude thirteen. (The larger the number, the fainter the object.) The asteroids following are listed according to their order of discovery.

Table 3. Some Large Asteroids

Asteroid	Diameter	Magnitude	Remarks
Ceres	428 miles	7	Visible in binoculars
Pallas	281 miles	9	Need six-inch reflector
Juno	150 miles	8	Closer to Earth than Pallas
Vesta	370 miles	6	Visible in small telescope
Astreae	112 miles	10	Use eight-inch to see it well

Observing Jupiter

The giant planet Jupiter has a diameter of about 87,000 miles. The planet will even show a disk in a pair of binoculars. With a small toy telescope such as a 2.4-inch refractor or a four-inch reflector, the planet is of a large enough size to see clearly. In these toy telescopes the two largest cloud belts located in the equatorial region can just be seen.

With a small telescope such as the six-inch reflector or the three-inch refractor, as many as six bands can be seen. With the moderate-size 'scopes and the larger 'scopes—both refractor and reflector—the amateur may join serious groups who study this gas giant.

Jupiter is the first of the gas giant planets and the largest planet in the solar system. We can never see the surface of the planet—when we look at it through our telescopes we see the structure of the cloud belts in the upper atmosphere. The atmosphere of Jupiter is very thick and the lower into the atmosphere the thicker and hotter it gets. The deep layers of the planet's atmosphere are so thick it is probably like melted plastic. You would be able to float in the Jovian atmosphere if you could survive there.

Jupiter's Belts

Jupiter moves into opposition with the Earth about once a year and this is the time to watch it. Professional astronomers have divided the planet up into as many as eighteen sections.

For the amateur astronomer this is not needed and I myself have remapped Jupiter for the amateur with ten sections. I'm sure that you will find that this ten-section division works better for you. For the amateur there is no need for countless subdivisions since his or her telescopes will not resolve them as well as the large observatory types.

I will list the divisions and cover what can be seen with a reflector of about eight or ten inches (see figure 6). Similar detail may also be seen with a four- or five-inch refractor.

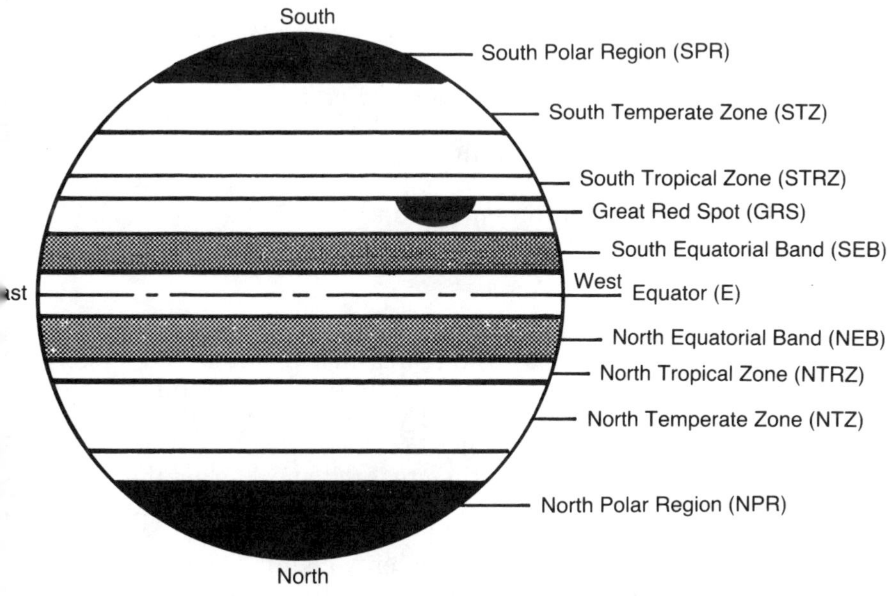

Figure 6. The Planet Jupiter. Location of main areas on Jupiter and how they appear through an eight-inch reflector during opposition.

Note: The Great Red Spot (GRS) although it is listed with the STRZ can be classified as its own area since it does move slightly. Theory has it that the GRS is above the cloud belt of the STRZ and not attached or part of this band.

South Polar Region (SPR) This is the area of the south pole of Jupiter. It is normally shown as a dark thick cloud band, but I have seen it composed of as many as four or five bands real close together. However at the top of the pole it does appear to be a solid brownish darkening. This could be due to the inability of many eight-inch to resolve them into individual bands. The bands in this area are brownish in color and a light yellow filter used with 300X or so will show them well. The stability of the atmosphere at the time will determine how many of these small bands you will be able to resolve. Try as much as 400X with an eight or ten in the area.

South Temperate Zone (STZ) This is just below the SPZ. Once again there are several bands that can be seen. In this area however they are somewhat further apart. An eight-inch reflector or four-inch refractor will show two or three of them. Two may be very close together and you many need an eight or a ten to resolve them. They look like one solid band at times and are often referred to as the double band by amateurs. Every once in a while you may see a storm which may look like a black spot or grayish line in contrast to the light brown bands in this region. I suggest that high power be used in order to get the most resolution.

South Tropical Zone (STRZ) This is one of the more interesting regions of Jupiter. It is in this area that the great red spot is visible. The STRZ consists basically of one dark thick band, brownish in color. In this band you will find much storm activity, and disturbances of all kinds.

I remember observing the great red spot during the opposition of 1975. The great red spot at this time was a blood red and it looked quite remarkable in contrast to the yellowish Jovian disk. Then on August twenty-eighth I noticed in the STRZ a very intense storm forming on the western limb of the planet. Several days later I looked at the storm area once again and I was amazed at what I saw. The storm had grown and was now moving across the planet toward the red spot. From night to night I followed its progress toward the red spot. This storm was a deep black in color and about as long as the red spot itself, but not as wide. It looked so out of place, this black cloud, that to see it in relation to the Jovian disk was quite startling.

For three months I followed the storm and it didn't seem to lose intensity at all. Usually the most intense Jovian storms live for only a month or less before losing energy. Finally on October 18, 1975 it reached the red spot. I observed this fantastic happening from 4h 00m UT to 6h 00m UT. In the telescope it seemed to graze just the top (southern) part of the red spot. From this time on the storm seemed to be sucked into the red spot (of course these observations were carried on over several weeks). Some of the storm however seemed to miss it, but the storm started to break apart. It seemed over the next few days to start to circle the red spot and as it did this the red spot started to fade. It turned from a deep red to a red, then to an orange. Shortly after this, Jupiter was low in the sky (since by this time it was moving close to conjunction with the sun). I tried looking for the red spot but could not find it. I thought at the time it might have been because the planet was moving further and further from the Earth. I eagerly awaited the next opposition which was to occur in 1976. The 1976 opposition occurred during November 15, 1976 and I searched the entire planet for several weeks for the great red spot.

I thought I saw a very faint yellow-orange patch in the STRZ one night, I then saw it several nights later. This must have been the great red spot, but it faded quite a bit and was barely visible in my ten-inch reflector at 300X. I wrote several groups of astronomers hoping that someone had seen the spot. I learned that many amateurs reported that the great red spot was not seen during the 1976 opposition. Some like myself did see what appeared to be the red spot but it was very faint.

Professional observatories soon announced that the great red spot had faded considerably and would just be visible in a 'scope of ten inches. They also recorded the great storm of 1975 and how it seemed to be swallowed up by the red spot. The question that has not been answered is could the storm have caused the great red spot to fade? The spot area is most likely a great cyclone which is swirling at a great speed and bringing up material from the lower layers of the Jovian atmosphere. The material that it brings up probably reacts with the gas in the upper atmosphere and produces a red color. If the great storm of 1975, as it entered the red spot area, disrupted it somewhat, it might have lost its ability to suck up a great amount of material from the lower atmosphere. This would cause it to fade and lose its distinct oval shape. This effect should be only temporary, and in time the red spot should once again become the most prominent feature in the cloud belts of Jupiter. As of the 1980 opposition the red spot has not been seen well at all. Amateurs have reported seeing an irregular oval patch a very light yellow-orange in color. It has never looked the same since the great storm of 1975 entered it.

The red spot has been known to fade from time to time and it has also been known to change from a light red to a deep red, a brick red, and so on. It was first observed in 1664 by Robert Hooke and between the date of its discovery to 1878 the color changed from a light red to a very deep blood red. In the late 1890s it faded considerably, but then once again became a prominent feature. The spot also has been noticed to wander slightly from its mean position, and one calculation showed that the spot wandered so much since its discovery that it made three revolutions around Jupiter independent of the planet's rotation. It will be interesting to see if the spot becomes easily visible once again in the near future.

South Equatorial Band (SEB) This is a very thick dark brown band and it is visible even in very small telescopes. In a six-inch reflector it appears as a straight, somewhat smooth band. You may see some irregularities in it with a telescope of this size if you look closely. A good number of storms and disturbances of all sizes can be seen on any given night. It is most interesting to watch the progress of a storm across the planet. Using an eight-inch reflector you can begin a serious study of this region and the band will no longer appear straight and smooth but very irregular with wisps of gas sweeping up into arches toward the STRZ and the equator. The appearance of this band as observed in an eight- or ten-inch will change from night to night. Some nights it may look darker or more broken up while other nights it may appear quite calm and smooth.

Using a twelve inch reflector you enter upon a whole new world. You will be able to study very small atmospheric changes and study in depth the atmospheric dynamics of the planet. During times of good seeing for planetary study (the calm hazy summer night) the view of this planet with an eight-inch reflector or a four-inch or larger refractor will steal your breath away. You may even notice some other colors like red-lined belts or greenish clouds and bright yellow oval spots. Most storms appear much darker than the normal Jovian cloud belts and will stand out very well with a four-inch refractor. The longer the focal length of the telescope the better the image you will get. A ten-inch f/7 or an eight-inch f/8 reflector works well. A reflector should have a focal length greater than sixty-five inches. A refractor should have a focal length of at least forty-five inches to get a good image. The four-inch f/15 refractor sixty-inch focal length is a fine choice for serious planetary studies of the planet Jupiter.

Equator (E) The equatorial region of the planet is usually much lighter than the rest of the disk. This could be because the pressure is much greater here and material is forced out to the equatorial bands. This can be seen through a ten-inch at 400X. Wisps and clumps of gas can be seen sliding out of the equatorial region. You will notice also that there are what appear to be segments of a broken line stretching from east to west across the disk at the equator. These are small storm systems and are seen quite often. I have included them in my drawing of Jupiter. Very rarely do you see a large storm in the equatorial region but it is filled with all kinds of strange atmospheric disturbances.

You should try using a green filter first on this area, then change to a dark or light yellow filter. Compare the two to see which one brings in more detail.

North Equatorial Band (NEB) This band is located just north of the equator. If your image is inverted this will be below the equatorial region, but if you are using a star diagonal it will be above the equatorial region. You can't miss this band since it is about as thick and dark as the SEB. It is brownish in color and many times it has been reported to be thicker than the SEB. A six-inch reflector will show it as a brownish band which will appear somewhat darker in the center area than at the ends. Just below this band is another brown band which is very close to the NEB. I don't think a six can resolve it and it may seem like one band. An eight under good seeing conditions will easily split them. With a ten-inch you can see quite a bit of space in between the two bands and see a great amount of disturbances. I have studied this area with a ten-inch f/7 reflector at about 450X and it seems as if sections of it are being torn apart. You can see bits of matter being drawn up into the SEB. I have also noted that as the matter is drawn toward the SEB at times it starts to take on a swirling motion. This in turn creates an oval brownish spot and must be similar to a hurricane on Earth but hundreds of times more intense. Such a storm was observed and recorded during the opposition of 1978 on March 3, 1980 at 4h 00m UT by myself and Milbank Observatory astronomer Rick Bria observing independently.

Once again you'll need high power with at least an eight-inch reflector to show the space between the bands. A four-inch refractor might give you a look at the area during moments of atmospheric stability. A yellow filter should be used to produce suitable contrast so that storm activity can be studied.

North Tropical Zone (NTRZ) There is very little here. Sometimes observers have reported seeing a thin light brown band. Here also you may see the strange-looking white spots. These are oval-shaped bright white spots which may be areas of intense high pressure. Try using moderate power in this area then slowly working your way up to the higher powers. While scanning the area if you should see something then check it out with about 300X; try using blue, green or yellow filters. It's very hard to say which filter will work best in this region and I suggest you try them all.

North Temperate Zone (NTZ) One faint band is seen here, brownish in color (although sometimes observers have reported a reddish tinge to a small section of it). This band is normally faint and might take an eight-inch to see it well. I have heard reports of this being a double band, that is, two bands that are close together. If this is a double band then they both are very thin and faint.

North Polar Region (NPR) Appears dark brownish or gray as you get closer to the pole. It is suggested that at least the southern part of this region may contain numerous bands very close together and unresolvable in amateur 'scopes.

Jupiter's Moons

Only four of the fifteen Jovian moons are visible with a telescope under twelve inches. These four moons are quite bright and can be seen very well with a six-inch reflector. They are often referred to as the Galilean moons, after Galileo who first saw them in the seventeenth century. A ten-inch reflector will show them as very bright star-like objects at 150X. They will be in line with the equatorial plane of Jupiter and will shift position from night to night. You can even detect some motions in their orbits after two hours, and this is why they are so interesting to watch.

A twelve-inch or larger will show them as very small disks at about 500X. There have been reports from amateurs using large scopes that they saw some surface markings on one of them.

Among these four, Io is the closest moon to Jupiter. It has a surface of frozen sulfur and sulfur dioxide with very active volcanos. We know this because of the recent pass by space probe Voyager I. Voyager I gave us information on the planet Jupiter and the Galilean moons that will take many years to analyze. Io is slightly smaller than our own moon and since it is so close to Jupiter its surface is kept constantly in motion. This is because of the great gravitational tidal forces exerted on it from the mother planet. If you were able to build a house on a stretch on land on Io and left for the weekend, when you returned the house would be gone. As a matter of fact the entire area would have changed beyond your recogition.

Europa is the Galilean moon second from Jupiter. It is the smallest of the four and is slightly smaller than Io. Voyager I sent back some excellent close views of Europa. It shows a smooth surface with very shallow wide streaks running across the planet in all directions. From a distance it looks almost like a reddish-pink marble.

Ganymede is the third moon from the mother planet. It is quite a large moon and is about the size of the planet Mercury.

It has a water-ice mantle and quite a few craters on its surface. It is very dark with basins similar to those on our moon, with many fractures in the surface that appear to be very young (astronomically speaking).

Callisto is the furthest out of the Galilean moons. It is about the size of our own moon and has a surface of ice and rock. Voyager I and II showed it to be covered by many craters.

Voyagers I and II gave us a fantastic view of Jupiter and its amazing moon system. I shall never forget the feeling that I had when I first saw the photographs that were released to the public—breathtaking!

The amateur astronomer who wants to do a serious study of the Galilean moons is limited to the observation of their orbital motions. This is much more exciting than it sounds. These Jovian moons exhibit some very interesting phenomena. These events can be observed even with a small telescope. There are three types of phenomena.

Transits occur when the sun, a Jovian moon, and Jupiter are in line. The sun, reflecting directly on the moon, will cast a shadow into the Jovian cloud belt. Observing one of these shadows will be quite interesting if you have never seen it before. The shadow on the cloud belts will look circular and will be black in color. The depth of the color of the shadow will depend on what Galiliean moon is casting it. If the shadow is being cast on an angle as seen from the Earth the shadow will look somewhat elongated. As the moon revolves around the planet so will the shadow move across the Jovian disk. The majority of these shadows fall somewhere between the SEB and STRZ.

If you were standing on top of the Jovian cloud belts and the shadow was covering your area you would see an eclipse of the sun. Transits occur quite frequently and a number of them can be observed during the weeks when Jupiter is high in the sky. I have seen several times when two transits occurred at once and two shadows were cast on the planet's cloud belt. Transits are easily visible since the black shadow stands out very well with the yellow white Jovian disk. Try a yellow filter—it should help to make the event more visible.

Occultations occur when one of the moons move behind the Jovian disk. This is an interesting event to behold since as the moon approaches the limb of the planet it becomes distorted in the atmosphere of Jupiter. Due to refraction phenomena the light of the moon will still be visible when it is physically behind the planet. It is so distorted however that it looks as if a flame was extending off the corner of the disk. It will then start to flicker wildly then slowly disappear.

Observing Saturn

The ringed planet stands out as the most beautiful sight in the entire solar system. Actually there are two other ringed planets in the solar system, Jupiter and Uranus. The ring systems of these planets are not visible in amateur-sized telescopes, but the Saturnian ring system is.

Observing the planet with a six-inch reflector will show the rings quite well and on nights of good seeing you should be able to see the Cassini division. This division is named after the famed astronomer Giovanni Cassini who first observed it in 1675. Cassini's division is a dark gap separating bands of the ring system. Through an eight-inch reflector it will look like a black line in between the rings.

A six-inch will show Saturn for the sake of just looking at it, but in order to really get a good look at the planet, an eight-inch reflector or better will be needed. Saturn at a distance of 900 million miles shows a small disk at even 200X. The combination of the ring system which is quite large and the planet however produce a good-sized image at medium powers.

The ring system is an amazing sight through an eight-inch, and you will be able to see that it is made up of a number of sections. When the rings are seen flat, face on, toward the Earth they look very wide and are a sight to behold. The rings are however very thin and when they are seen edge on from the Earth they seem to vanish. As of May of 1980 the rings can be seen with an eight-inch as a thin bright line running through the planet. The rings did disappear for a while early in 1980. They are now starting to once again turn face on to the Earth and by several more oppositions they should be very prominent (through the eighties and until 1995).

Although you won't be able to see a great deal of detail on

the atmosphere of the planet what you will see will interest you. An eight will show some cloud belts around the equator and some faint darkening around the poles. An eight-inch will also show you the shadow of the ring being cast across the disk of the planet. When the rings are thin or edge on as seen from the Earth the rings shadow is very visible. Through an eight it looks like a grey-black pencil line stretching across the equatorial region of the planet. If you look well you will see that it appears to divide the planet into two sections. This gives a strong impression that the planet is cut in half. It is an interesting phenomenon to see, but you'll need about 300X and a minimum of an eight-inch reflector or a five-inch refractor. Through a ten- or a twelve-inch the detail becomes very clear and pronounced. The equatorial clouds may be mapped, and changes in them will be easily noted.

Near Saturn's equatorial region unusual white oval patches can occasionally be seen. These areas are similar to the Jovian yellow-white elongated markings which are also seen near the equator. The white oval features in the cloud structure of Saturn, although more defined in shape, are much fainter. You will need at least a ten-inch reflector or a six-inch refractor to get a good look at them.

The upper atmosphere of Saturn is not as active as the Jovian atmosphere, but from time to time you may observe some very interesting phenomena. Saturn's image seems to hold very well under high magnification and you may want to try as much as 500X with an eight-incher. Once again a telescope with a focal length of sixty inches or better will work the best and give you a sharp image. Although as of this writing Saturn is known to have fifteen moons, only five are of much interest to the amateur astronomer. These five moons can easily be seen with a small telescope on a clear moonless night. The larger the telescope you use the better they will stand out. An eight-inch reflector will show them to be very bright and distinct at about 125X.

From night to night you will see them shift position in relations to their mother planet. Some nights, only two or three may be visible since the others may be behind Saturn. Below, the five moons that can be observed without trouble by the amateur are listed and some basic information is given.

Titan is the brightest of the moons and is located at about the same inclination as the ring system. It has an apparent magnitude of about eight and is quite visible even in a 2.4-inch refractor toy telescope. It should have a whitish color with a red tinge. Titan is the largest moon on the Saturnian system and is the only moon known to have a substantial atmosphere. It is larger than any of the Jovian moons. With a diameter of 3,600 miles it almost classifies as a planet. Titan orbits Saturn about every fifteen days.

Rhea is the second-largest moon in the system and the second-brightest. A six-inch reflector will show it very will since it has an apparent magnitude of nine (about 2.5 times fainter than Titan). Rhea has a diameter of 980 miles and revolves around Saturn every 4.5 days. Rhea lies very close to the orbital inclination of Titan, and many times the two can be seen lined up, with Rhea closest to the planet.

Dione will be close to the disk of the planet and will be somewhat faint since the glare of the Saturn will almost overwhelm it. Dione still is easily seen with an eight-inch reflector. At times it will be very close to the rings and at this time it will be somewhat difficult to see. A six-inch reflector will show it well when it is at its greatest separation from the mother planet. Dione shines at an apparent magnitude of about 10.2, but seems such fainter when it is at its closest to Saturn. Dione was discovered in 1684 by Cassini and has a diameter of about 500 miles. The moon takes 2.6 days to revolve around Saturn.

Tethys is a little larger than Dione; with a diameter of 625 miles it appears to be about the same brightness. This is because it is closer to Saturn and is overwhelmed by the bright disk and ring system. Dione and Tethys, when seen together, look like twins since their visual magnitude is about the same. Tethys takes only 1.9 days to revolve around the planet Saturn.

It will be hard to get a good look at these moons with the rings face on or open. The ring system when seen in this position can be very bright. The best time to photograph or observe these two moons is when the rings are edge on and invisible from the Earth. This will make the moons stand out quite well on a clear moonless night since you won't have the glare of the rings to compete with.

Iapetus is the second most distant moon from Saturn. It

was discovered in 1671 by Cassini. Iapetus shines at an apparent magnitude of eleven and is 800 miles in diameter. You might have a little trouble seeing it with a six-inch if you live in an area where the light pollution is bad. If you live out in the country and enjoy the pleasure of dark skies you will have no trouble seeing it with a six-inch reflector. An eight or ten will show it to be very bright and you will be able to follow it from week to week as it revolves around Saturn. In different parts of its orbital positions it may vary slightly in brightness. Although Iapetus is larger than Dione or Tethys it is fainter. This is because it is believed to have a dark rock surface which reflects only a small percentage of light. It is also thought that some areas may be covered with bright snow-like material and this might account for the fluctuation of brightness as it orbits Saturn.

Observing Uranus

For the amateur astronomer with a telescope under sixteen inches Uranus offers little. Many who observe it find it, look at it for a while, then turn the 'scope to more interesting objects. A six-inch reflector will barely show the small disk. An eight-inch will show it a little better, but you'll need at least a ten-inch reflector to see it well. Using an eight-inch at 250X, Uranus displays a small blue-white disk about the size of a pea at ten feet, as seen with your eye. A ten or twelve at 400X will show a good-sized disk, but in comparison to some of the other planets it will still be very small.

Amateurs with the large twelve-inchers or fourteen-inchers have reported a faint equatorial band and some polar darkening. William Herschel, who discovered it in 1781, also reported some type of marking across the equator. Uranus shines in the sky at a magnitude of about six which makes it just visible to the naked eye. The planet can be seen as a star-like object with small toy telescopes, and binoculars. It is important to understand what you are seeing when you observe Uranus since it will help you appreciate it more.

Uranus has an equatorial diameter of 33,000 miles, which makes it quite a large planet. It displays such a small disk because it is one billion, seven hundred and eighty million miles from the sun. It is about 1.7 billion miles from the Earth. Uranus is one of the gas giants like Saturn and Jupiter. We only see the upper atmosphere of a planet which is composed of methane, ammonia, hydrogen, and some helium. Uranus has five moons which require a large telescope of fourteen inches or so to even spot. Three of these moons can be photographed with a twelve-inch reflector with a very long exposure. These moons are called Ariel, Titania and Oberon.

To locate Uranus you might want to use your setting circles—you can get the coordinates from *Astronomy* magazine. The best time to observe the planet is during an opposition since at this time it will reach its highest point in the sky at around midnight. If you don't have setting circles, *Astronomy* magazine usually will publish a chart and star map showing you where to find it in relation to the stars in the area. Uranus is quite easy to find, but use at least 200X to look for it or you may pass it by.

Observing Neptune

Needless to say, the disk of Neptune as seen through amateur telescopes is even smaller than Uranus. You may see its tiny green-blue disk at about 300X with an eight-inch reflector under good conditions. A large telescope is badly needed for observation of this planet. Astronomers think that the atmospheric composition is very similar to Uranus. Neptune is slightly smaller than Neptune, having an equatorial diameter of about 30,750 miles. Neptune is at a distance of almost three billion miles from the Earth (the actual figure is 2,794,100,000 miles).

Neptune has two known satellites. They are Triton and Nerid. Triton is thought to be the largest moon in the solar system with a diameter of 3,800 miles. Although this moon is within the photographic reaches of a large amateur telescope, it is doubtful it can be photographed with amateur equipment. This is because the separation from the disk of Neptune to Triton will be very small and the image of an overexposed Neptune will cover it.

If you wish to observe Neptune you'll have to use your setting circles to find it since it is at a magnitude of about eight, almost ten times fainter visually than Uranus. You should wait for an opposition (which occurs once a year) and try to pick a night with the best possible conditions. It will take 300X or so to show the disk well enough so that you don't miss it and pass it by while looking for it. Magazines like *Astronomy* or *Sky and Telescope* will publish detailed maps on where to find Neptune during an opposition.

Amateur astronomers with small and medium 'scopes pride themselves when they talk about "observing Neptune last night." It is worthwhile to find, but a very hard object to observe.

Observing Pluto

Unless you live at least one hundred miles from a major city forget about looking for it. Although visually it can be seen faintly with a ten-inch reflector you will need very dark skies and accurate maps and circles to find it. When you do think you have found it you won't be sure since it will look like a very faint star. The only way in which you can identify Pluto is to take a series of photographs about two weeks apart. Pluto, being much closer than the background stars, will change position slightly in relation to them. I have great respect for amateurs who have photographed this tiny planet with modest equipment. We (our local group) tried to photograph it during the 1980 opposition but in our area the skyglow and light pollution was too severe. The image of Pluto was lost in the skyglow on the photographic plate. If you have dark skies and good photographic equipment I say, go for it. If you succeed you will be one of the few amateurs who have. Data for finding Pluto can be obtained through a number of magazines for the amateur astronomer.

Observation of Special Phenomena

This is quite a broad category but I will limit it to a smaller number of spectacular events. We will be talking about events which occur for only a short time and are a one time happening.

Lunar Eclipses

A lunar eclipse (see figure 7) occurs when the moon moves into the shadow of the Earth. At this time the Earth will be between the sun and the moon and will cast a shadow in the shape of a cone into the space behind it. This shadow is composed of two sections—the penumbra or light shadow and the umbra or dark shadow. A total eclipse of the moon can only occur during a full moon and as the moon moves into the penumbra shadow it will lose a small amount of its brilliance.

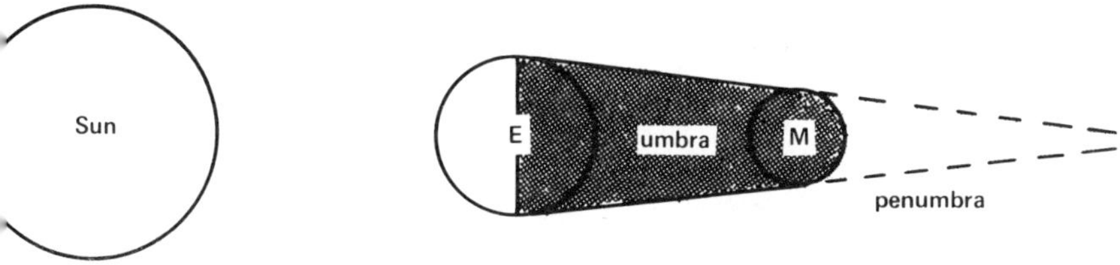

Figure 7. A Lunar Eclipse. When the moon passes completely into the *umbra*, the eclipse is total.

Lunar eclipses occur once or twice a year, in general. See any of the astronomy magazines for dates and times.

While the moon is totally within the Earth's shadow it can still be seen faintly. No two total eclipses of the moon are the same. At the time of full totality the moon will display an unusual color which will vary from eclipse to eclipse. The color during a total eclipse has ranged from a grey-white to a dull, copper, or deep red. The colors however stay within the range of shades of gray and shades of red. Amateur astronomers record the color to compare with past and future eclipses.

The color variation is due to the sun's light refracting through the atmosphere of the Earth, reaching the moon, then reflecting from the lunar surface to the Earth.

At times, the color of the moon in eclipse will be bright enough to see the lunar mare with the naked eye. Most of the time a telescope will be needed to see any of the surface detail since the light will be very faint.

You can use even a very small telescope to observe the eclipse and watch how the shadow of the Earth slowly moves across the lunar surface. At times the lunar features will throw interesting and sometimes very strange-looking shadows as that section of the moon moves into the umbra of the Earth's shadow. You might want to observe this shadowing phenomena at about 150X, but watch the eclipse itself at low powers, somewhere between 25–60X.

Eclipse totality time will differ with every eclipse, due to the very complex motions of the Earth and moon as they revolve around the sun. As the moon moves out of the Earth's umbral shadow a wave of appearance will begin and more and more of the moon will once again become fully visible until the moon is once again full.

What we talked about is the total lunar eclipse which occurs when the moon moves completly into the umbra or dark shadow of the Earth. There is another type of eclipse called a partial eclipse. This occurs when only part of the moon moves into the umbra of the Earth. As seen from the Earth only a part of the moon will become darkened; the percentage that does become engulfed in the Earth's shadow will depend on how much of the moon moves into the shadow.

A third type of eclipse is called a penumbral eclipse and this occurs when the moon moves only into the penumbra or light shadow of the Earth. As seen from the Earth the full moon will slightly become dimmer; unless you have very keen sight you will hardly know the difference. Penumbral eclipses are not very interesting and at times not worth seeing.

Solar Eclipse

When the disk of the moon moves in front of the sun, occulting it, we have a solar eclipse. A solar eclipse can only occur at new moon since at this time the moon will be between the Earth and the sun. Strangely enough, is the fact that the disk of the moon just about covers the disk of the sun, since they are the same apparent size as seen from the Earth. This is what makes a solar eclipse a spectacular sight.

The total eclipse can only be seen in a small area when it occurs. This is because the shadow cast by the moon onto the Earth is very small. The dark shadow cast onto the Earth is once again called the umbra, while the lighter shadow around it is called the penumbra. It is in the area of the umbra that a total eclipse of the sun will occur. Outside, but close to the umbra area, observers will see a partial eclipse. This umbra shadow covers an area about 167 miles wide; the longest totality recorded was seven minutes and fourteen seconds on June 30, 1973.

Figure 8. A Solar Eclipse.

The sun can also display a partial eclipse and as seen from the Earth a certain percentage of the sun will be blacked out. Many amateurs will travel to locations thousands of miles away to be in the umbral area during the time of a total solar eclipse.

To observe a solar eclipse you will have to follow the same precautions as in solar observing. There are two safe methods: The first is that Solar Skreen full-aperture filter by Roger Tuthill Inc. The second is projection onto a screen. If you have several people with you then the projection screen may be a better choice. With this method all can watch in perfect safety and not miss any of the eclipse. Photographs may also be taken of the image on the screen for a record of the event. For direct viewing and high resolution photography, the full-aperture solar filter is the better choice. It is also the better choice if you are watching the eclipse by yourself since it will produce a sharper image.

To observe a solar eclipse any telescope on a mount will do. Even the small toy telescopes will show the sun in eclipse well. I suggest that you still use the equatorial mount since it will allow you to track the Sun without too much trouble.

Solar eclipses are more frequent than lunar eclipes, but the solar eclipse will only be seen in a very restricted area. It is a major observing event for the amateur astronomer who prepares to photograph, study, and record a total solar eclipse of the sun.

Table 4. Future Solar Eclipses*

Date	Type	Location
Dec. 4, 1984	Partial	United States
Oct. 3, 1986	Total	Canada, North Atlantic ocean
Mar. 29, 1987	Total	Mid. Atlantic ocean
May 10, 1994	Partial	Canada, United States

*Visible in North America

If the eclipse is to occur over water, many amateurs take their equipment on special chartered ship cruises which not only offer a flat deck for observation, but also several days of lectures, slide shows, and so on.

One such travel agency that offers this service is called Solar Eclipse International at the Granada Hills Travel Service, 16800 Devonshire Blvd., Granada Hills, California. They will be pleased to send you further information if you desire to go eclipse chasing one of these days.

Observing Comets

Discovering a comet is a rare honor, but it has been done many times by amateur astronomers. A comet is most visible when it is closest to the sun. Comets should be looked for before sunset and sunrise.

Wide-angle high-power binoculars (25 × 125) work best but can be very expensive. The Newtonian RFT with a focal length of f/5 or less also works well. The angular size of a comet in the sky is great but it can be very faint, so this is why you need the wide angle equipment for low magnificaion and brighter images. Every amateur is waiting for the return of Halley's Comet in 1986. The last great bright comet was Comet West in March of 1975.

Observing Meteor Showers

There are several good showers a year and they can be quite interesting events. All you need is your eyes, a star chart, a reclining chair and perhaps a pair of binoculars to follow the more long-lasting ones.

The best way to observe a shower is with at least three people. A field with a unrestricted view with dark skies must be chosen. Meteor showers are associated with certain constellations which are their radiant points. Two observers lying on reclining chairs point themselves toward the constellation radiant point. They are spaced so that the entire constellation plus an area on both sides of it is covered. The third observers sits at a table or sits in a chair with star charts, pad and pencil and perhaps a tape recorder. As a meteor becomes visible, the observers will call out its location, direction, brightness (magnitude), and color. Also remarks are given as to how much of a tail was seen its duration, and so on. The observer with the pad and pencil and star charts will plot the meteor on the maps and list the information about it in a log or pad. The observers in the chairs can have a table between them with one or two pairs of binoculars on it to check out some of the more interesting ones.

Meteors can vary a great deal in brightness and at times they may be just visible to the naked eye. Other times they may light up the entire area. Colors will range from red to white to yellow and perhaps green.

Table 5. Yearly Meteor Showers

Date	Constellation	Meteors per hour
April 22	Lyra	10
July 30	Auquaris	30
August 12	Perseus	50–70
November 8	Taurus	15
December 14	Gemini	45–55

The dates above are when the shower is at its peak. At this time the maximun number per hour can be seen. The shower can be seen as early as two days before and as late as two days after the peak date, but the number of meteors seen will be greatly reduced.

Observing Artificial Satellites

In the past twenty-three years a number of countries on the Earth have launched artificial satellites into orbit around the Earth. Most of them orbit the Earth in about ninety minutes and can be seen several hours after sunset and several hours before sunrise. There are now somewhere between 2,000 and 3,000 man made objects circling the Earth. Most however are space junk. Space junk is the remains of a booster rocket, an old turned-off weather or spy satellite, and so on.

Go out to a field on a clear night away from the street light and look up towards the night sky. Soon you will see a star-like object moving across the sky at a moderate speed with a slight bobbing motion. Satellites are visible due to reflected sunlight and the size, and distance from the observer will depend on how bright it gets.

I remember watching Skylab as it passed almost directly over the New York-Connecticut area. Skylab was quite large and it was much brighter than the brightest stars in the sky. I estimated its brightness at times at almost that of the planet Venus. When it was far from my area, however, it was very faint.

If you go out any clear night at the times listed above you will see at least four to six of these man made objects. At times they might pulsate and may flare up and become very bright for a few seconds then fade out to the limit of visibility. This is due to the fact that some of them may be tumbling in space and might be reflecting the sun's light unevenly.

A good pair of binoculars will come in handy to help you follow the object in its motions through space. Wide-angle equipment works the best since satellites travel too fast for any telescope to follow them. The field of view through a typical telescope is so small that the satellite would streak by in a second or so. This is because most astronomical telescopes produce too much magnification to follow their movements.

Most of the satellites that you see will not be named, but an occasional named satellite may pass over your area. Many planetariums across the country carry information as to satellites will be seen, and from what country they were launched. Satellites are more or less a sideline of the amateur astronomer, but they are a great deal of fun to watch. There is also the chance that you may see the disintegration of one since they are all slowly falling back to Earth. This will be a sight that you will never forget. The fireball produced by a satellite burning up on reentry will become brighter than the full moon and it will light up the entire area.

IV

OBSERVING MULTIPLE-STAR SYSTEMS

We now leave the solar system and turn our telescopes to the distant stars. Most stars occur in groups. When some of their members are spaced light years apart, the groups are called loose associations. The sun and the nearby Centauri system make up the local family of stars to which the sun belongs. Other types of groups are spaced very closely. These systems are called binary stars if there are two members and multiple systems if they contain more than two stars. Some of these systems are actually physically close together and revolve around each other at a common center of gravity. Some of these systems are only close as seen from the Earth since they are in the same line of sight. They are called visual binary or multiple systems and are also of interest to the amateur astronomer.

In some of these multiple systems the stars are spaced far enough apart to be seen with the naked eye or binoculars. Others will need a telescope and sometimes very high powers. Some of the closer binary stars will, for example, under low powers, look like one star, but under high power will split into two stars.

I shall cover the more interesting multiple and binary star systems and where to find them. I shall also cover what equipment to use to split them into their components. The separation between the members of the closer systems is given in "arc seconds" or "minutes." The diameter of the moon in angular size is about thirty minutes. The angular size of the disk of

Jupiter averages about 35–40 arc seconds.

Multiple Stars

Polaris

A good place to start is the North Star. Polaris is an easy double star but you will need at least a six-inch reflector to see the companion well. Polaris itself is a second-magnitude star and the companion can be seen eighteen seconds from it. The companion seems to be very blue color and is located almost directly above the star. The companion is a ninth magnitude star and is somewhat subdued in the light of Polaris, which is very bright in a six-inch. Use about 150X to get the greatest distance from the primary star. A smaller telescope may not show it since it looks much fainter than it is, being partially drowned out in the light of Polaris.

Polaris can of course be found in the handle of the little dipper (Ursa Minor).

Mira

This is an unusual star that changes its brightness considerably. At times it can be as bright as a third-magnitude star and at others times fade to a fifth-magnitude star. The companion is a tenth magnitude star and separated from the primary by only 0.8 seconds of arc. This companion actually orbits Mira and when it passes in front of it it makes Mira fade since it occults some of the stars light.

You will need at least a twelve-inch on a good clear stable night to separate these two stars. In theory an eight-inch reflector or a six-inch refractor should do it, but our atmosphere decreases the resolution power of our telescopes quite a bit. If you were, lets say, observing from the Moon then you would be able to split Mira into two stars with a smaller 'scope than a twelve-incher. Mira is located in the constellation of Cetus (the whale) about three degrees south of the equator (0 declination). If you do have a larger 'scope you must use the highest power possible for the size of your telescope. Mira is an eclipsing binary star.

Algol

This is another eclipsing variable star which changes its brightness quite a bit due to the orbiting secondary companion star passing in front of it. The name Algol means "winking eye" and was named by the ancients who noticed the fluctuations in its brightness. It will change in brightness from a second-magnitude star to a third-magnitude star. The companion star has a magnitude of 10.5 which makes it almost 100 times fainter than Algol. It is separated from Algol by over one minute. Algol is in fact a multiple system containing three members. The star that eclipses Algol is not seen, but the third companion is. You will need a six-inch telescope to see the visible companion since it is very faint. Its separation is quite far (about one and a half times the diameter of the disk of Jupiter from the primary star). Algol can be an interesting project for school since you can plot the changes in its brightness on a graph. It moves from its maximun brightness to its minimum brightness in just under three days. Algol is located in the right extension of Perseus and when it is at its brightest it is one of the brightest stars in the constellation and can't be missed.

Aldebaran

Aldebaran is the brightest star in Taurus (the bull). Aldebaran is an orange giant and easy visible as one of the most beautiful stars in the winter sky. This star is in fact a double star with an eleventh-magnitude companion located thirty-one seconds from the primary star (Aldebaran). You should use at least an eight-inch telescope to see it. A four-inch refractor shows it very well and the color of both stars is a beautiful orange and yellow. In order to see the companion wait for a clear dark night and go far away from the street lights and town lights.

Rigel

Rigel is a bright-blue first-magnitude star located in the eastern corner of the constellation Orion (the hunter). Rigel is also a double star and the companion can be seen very well in a six-inch reflector or a three-inch refractor at 150X. The companion is separated from Rigel by 9.2 seconds of arc and is of the seventh magnitude. You may have to look very hard for it since Rigel is so bright it might be partially lost in the star's glare. Once you do find it, however, it stands out very well and you will have no trouble spotting it again. It looks blue-green in color in a small refractor and is a beautiful sight.

Sirius

Sirius is located just east of Orion. Sirius is the brightest star in the sky and marks the constellation of Canis Major, or the big dog. In mythology, Canis Major is the faithful hunting dog of the mighty hunter Orion. He faithfully folows his master across the heavens. Sirius is for this reason also called the Dog Star. Sirius is a double star and the companion (Sirius B) is a white dwarf located very close to Sirius. Sirius B, although it has a magnitude of about 9, can only be seen when it is at its farthest point in its orbit from Sirius. This occurs every sixty years and will not occur for at least another forty years. Even at this time, since Sirius is so bright, B is lost in its glare and you will need a large amateur-sized telescope to even get a glimpse of it.

Sirius B has been the subject of a great debate over the past several years. It seems a primitive tribe in Africa knew of its existence centuries before it was discovered in the Western world. This seems highly unusual since you would need quite a large telescope to see it, yet this tribe also knew of its fifty year orbit. This has been called the Sirius mystery and is supposed to be evidence of the ancient astronaut theories. There are a good number of books out, both pro and con, about the "Sirius mystery" and it is not the purpose of this book to discuss it any further.

To photograph Sirius B is a great feat for the amateur astrophotographer.

Castor

Castor is the brightest star of Gemini (the twins) and is a double star. The companion is of the second magnitude and will need a moderate-to-large telescope to see it. The separation between them varies since this is a binary star. The distance between the two seems to be closing.

Regulus

The lead star of Leo (the lion) is an easy double to spot. Regulus itself is a first-magnitude star while the companion is about an eighth-magnitude star. Is is a gaping three minutes of arc from the primary star. It is visible easily in a three-inch refractor.

Mizar and Alcor

These two stars are located on the handle of the Big Dipper. To the naked eye they appear as two stars very close together. It will take a clear night and very good eyesight to see them as two stars. They are both very bright at about the second magnitude. As seen through the finder of your telescope they will appear as two separate stars with good distance between them. Mizar and Alcor are apparent double stars which means they only seem close together because they are in the same line of sight as seen from the Earth. Using a six-inch telescope you will see a very interesting thing—Mizar itself has a companion. Mizar B, as it is called, has an apparent magnitude of four and is fourteen seconds of arc from Mizar. It can be clearly seen with a small amateur telescope at about 100X.

Alberio (B Cygni)

Located in the "tail" of Cygnus (the swan). This is a very beautiful double star. It can be seen with a small telescope even at 50X and is thirty-five seconds from the main star Alberio.

Alberio itself is a third-magnitude star with a companion of the fifth magnitude. I also look at this double every once in a while because it is very colorful. Alberio is a yellow-orange star while the companion is blue-green. A three- or four-inch refractor will show it best since, with the reflectors, the star's image may be somewhat flaring or spiked slightly and this will take away some of its beauty. Refractors will show nice sharp round images and will suffer less from atmospheric turbulence. B Cygni is an object that must be shown to friends during star parties. The general comment: "I never knew the stars had color like that."

Epsilon Lyrae

This outstanding multiple-star system is located not far from the bright star Vega in Lyra. With the naked eye you can see two fourth-magnitude stars very close together. The separation between these two stars is just over three minutes of arc (about one tenth the diameter of the full moon). Both stars are white in color but the most amazing thing about these two stars is that they are in turn both double stars themselves. This is a multiple system with four members. The star with the widest separation from its companion is called E1. The E1 companion is a sixth-magnitude star and is separated from the primary star by a tiny 2.8 seconds of arc. E2, as the second one is called, has a companion with magnitude of also about six. It is separated from the primary by 2.3 seconds of arc.

This system is called the double-double by amateur astronomers because that's exactly what a quadruple star system is. To observe the double-double is easy. A six-inch reflector will split E1 and show both stars but unless you live in a location where the atmosphere is very stable and calm it will not completely split E2 into two stars. A three-inch refractor may split both stars at times of extreme atmospheric stability, but in order to see them well you will need a minimum of an eight-inch reflector. A four-inch refractor also works quite well and the refractor may be the better choice when observing the double-double. Since the components are very close together you will have to use high power to obtain as much resolution as you can. Start out with about 150X and slowly work your way up to about 300X to see what magnification will work best for the conditions of the night.

I have seen the double-double through many different types of 'scopes, but the best was through a Unitron four-inch refractor at 250X. I have never seen so much space between the two stars and their companions through any other 'scope of amateur size. That night we had a ten-inch reflector and the four-inch refractor and we used both of them on the double-double. The four-inch refractor at 250X showed the quadruple star system much more clearly than the larger reflector at the same magnification. It is a known fact that refractors do work better for the observation and photography of binary and multiple-star systems.

It is impossible to cover all the multiple star systems in the sky, but I have covered the more interesting, brighter, ones. They are very easy to find with the aid of a star map. Observing double stars is a good way to conditon your eye to pick out faint details. Double stars are for the most part very beautiful since some will show companions with fantastic color contrast.

Observing Star Clusters

Star clusters belong to the category known as deep-sky objects. There are two basic types of star clusters, open and globular. Open clusters are loosely packed groups of stars which to the naked eye may look like a hazy patch in the sky. Through even a small telescope the small hazy patch will resolve into many stars. Open star clusters are quite large in size and are best seen in the reflector-type telescope with short focal lengths. A six- or eight-inch f/5 or f/4 is a good choice since these 'scopes provide wide field, bright images, and low magnification. Open clusters are for the most part young stars that are born in groups out of nebulous gas. In some instances the nebula that created them can still be seen enveloping some of the larger and brighter stars in the group. The younger groups are very close together and contain hot blue-white stars. The older open clusters are very wide and sparse and contain middle-aged yellow and yellow-orange stars. Open clusters are often referred to as galactic clusters since they are found within our own galaxy.

Globular clusters are hundreds of thousands of stars tightly packed into huge balls. Through the telescope they indeed look like balls of stars. Globular clusters are older clusters, possibly created shortly after the formation of the galaxy. They are located outside the galaxy and over 100 of them form a sort of halo around the outer central area of the galaxy. Some of them seem very small and others are large as seen through the telescope. Actually they are all very large, it's just that they are very far away from the Earth.

Most of the globulars can be resolved by amateur telescopes, though some may just look like a fuzzy ball of light. The larger the telescope the better you will see them. An eight-inch reflector at f/8 works very well and will completely or partially resolve 80 percent of them. A four-inch refractor will also work well, but in any case use about 50–150X.

Globulars and open clusters are classified by a letter and number and for the amateur there is a chart in publication which has a listing of all deep sky objects easily visible in amateur-size telescopes. The listing is called the Messier catalog after the French astronomer and comet hunter Charles Messier. Messier who lived in the latter part of the eighteenth and early part of the nineteenth century was an avid comet hunter. Often seeing certain faint blobs of light through his telescope he began cataloging these fuzzy patches of light and listing their locations. The purpose of this was so that they would not be mistaken for comets by himself or other astronomers.

Messier's list includes 107 objects, including star clusters, nebulae and galaxies. This list does not contain all the deep-sky objects, but it does contain all the brighter and more interesting ones. Let's cover a few open clusters first and I shall discuss what they look like where they are located and how to observe them.

Object M-6

A large cluster located in Scorpio, it measures almost thirty minutes of arc (the diameter of the full moon). It is a very beautiful sight with a high concentration of stars toward the center. Use very low power in the RFT-type telescope, 20X is a good choice. A long-focal-length telescope will not work well on M-6 since it will provide too much magnification. If you use too much power the object will not be contained in the field of view and you will lose the effect.

Object M-7

This is a very large fifth-magnitude cluster which is also located in Scorpio. This angular diameter is almost two full moon. It is easily visible in a very small pair of binoculars and is also visible to the eye as a faint diffuse patch.

Object M-11

A small dense open cluster of the sixth magnitude, it measures 12 minutes (half the diameter of the full moon). It can be easily seen through a six-inch reflector and about 60X seems to work well. M-11 lies in Scutum.

Object M-16

A giant cluster in Serpens, M-16 is a small seven minutes in apparent size and of the eighth magnitude. This is a small and faint but interesting open cluster. In a six-inch reflector it may look like nothing much, but in an eight or ten-inch at about 100X it is a grand sight.

Object M-21

Located in the constellation of Sagittarius, M-21 shines at the seventh magnitude and has a size of twelve arc minutes. It is a beautiful and compact cluster whose center stars shine like ice crystals.

Object M-23

Also located in Sagittarius. It is twice the size of M-21 yet is just as faint. This is because it is much more loosely packed. It may just look like a faint conglomeration of stars, but use low powers anyway to get the best look.

Object M-24

Once again we look into Sagittarius for this sixth-magnitude cluster. M-24 is a tiny four minute of arc in size. A six-inch will show it well at about 80X, but an eight will give you a breathtaking view of a very compact open cluster.

Object M-29

Located in Cygnus the swan, it is seven minutes in size and shines at magnitude seven. It is somewhat loosely packed but worth a look.

Object M-34

A great object to look at, it is located in the constellation of Perseus and has the apparent size of the full moon. It shines in the sky at magnitude six and can be seen even in very small telescopes. It is also a great sight in a pair of binoculars. Use low power to observe this one; it contains many blue-white stars and you may count as many as forty or fifty with a six-inch reflector at 50X.

Object M-35

This is a beautiful cluster located in Gemini. M-35 is an interesting object also. It contains many colorful stars and the central part of it is very densely packed with bright stars. The stars' density thins out considerably toward the ends. Like M-34, M-35 covers an area in size about that of the full moon and has an apparent magnitude of six.

Object M-36

This open cluster in Auriga shines in the sky at magnitude six and has an apparent size of sixteen minutes of arc. It is a moderate-size cluster with about twenty-five main members. An eight-inch will show it very well at about 50X.

Object M-37

This also lies in the constellation of Auriga. It is loosely packed, has an apparent magnitude of six and is twenty-four minutes in size. It is an interesting sight in a small telescope.

Object M-44

Also called the Beehive cluster this is a very large object over three times the full moon in area (90 arc-minutes). It shines in the sky at magnitude four and is easily visible to the unaided eye as a faint hazy patch on a clear night. For this cluster you will need a wide-angle telescope; a six-inch f/4 at about 15X woks very well. There are about forty to fifty bright members and they look like little diamonds though the telescope. This is a very beautiful cluster and is very young. The stars are blue-white. Look for this jewel of the heaven in Cancer (the crab). It also will give you a good showing in even a small pair of binoculars.

Object M-45

Also called the Pleiades or the Seven Sisters. It is a winter cluster located in Taurus the bull. It can be easily seen by the naked eye as a group of glimmering giant stars in the night sky. With your naked eye you will be able to count six or seven members of this cluster. Through a pair of binoculars you may count twenty or thirty. To observe it with a telescope you will need very low power. This is because this cluster is so large that a more powerful telescope will provide so much magnification that the cluster will spill out of its field of view. This will allow you to observe only a part of the object and not the entire object, and a great amount of its beauty will be lost. A six- or eight-inch f/4 at no more than 30X will work well, but the lower the magnification the better the view you will get. M-45 is 120 arc-minutes (4 times the diameter of the full moon).

M-45 is a very young cluster—about twenty million years old. If you reduce the universe's age to the lifespan of a human then the stars in M-45 would be only several hours old. The cluster contains many stars which are a hot-blue to blue-white to white in color. If you look closely enough on a clear dark night away from city lights you may see faint traces of the nebula that created this cluster circling the brighter members. This nebula can be photographed through an RFT with blue-sensitive film. M-45 lies in our galaxy at a distance of 415 light-years from the Earth. The number of members that you will be able to count will depend on the aperture of your telescope and the focal number. With an eight-inch f/5 for example you should be able to count at least 150 members. The Pleiades is a sight you will want to show to other people since it fits the child's rhyme, "like a diamond in the sky."

Object M-52

A small, faint, but interesting open cluster in Cassiopeia. M-52 is only thirteen minutes in diameter and is of the eighth magnitude which makes it faint for an open cluster. It contains few bright stars but is very compact toward the center of the cluster. A six-inch will show it, but you may need an eight-inch to see it well. You might want to use about from 50–100X on this one since you will need moderate magnification to resolve the center area.

Object M-67

An interesting little cluster in Cancer, M-67 has an apparent magnitude of about seven and is about eighteen minutes of arc in diameter. It contains about thirty members which are evenly spread apart in relation to each other. It will look quite good in a six-inch reflector with a focal length of f/5 or so at about 50X.

The last open cluster that I would like to cover does not have an M-number. This is the double cluster in Perseus and is actually two open clusters joined together in a linear fashion as seen from the Earth. It is strange that Messier never recorded it in his listing since it is a large enough object to be seen in a good pair of binoculars. The double cluster has a size of about sixty minutes of arc and both clusters are about the same brightness. Their actual separation from each other is about fifty light years and both are very young—about fifteen million years old. They contain very hot blue stars, but some red giants can also be seen. Both clusters are very concentrated toward the centers and you can see very slight traces of the nebula that created them. The double cluster lies at a distance of 7,000 light years, which is much farther than most of the other clusters that were covered earlier. This object is a fantastic sight in even a telescope of six inches and you may count so many members that you might need some type of calculating unit to obtain an accurate count. You must use low power, as with most open clusters, since they are large in angular size.

Observing Globular Clusters

The second type of star clusters are globular clusters. I myself find them one of the most fascinating of all deep-sky objects and most of them can be seen well in a six-inch reflector. If you have never seen one through an amateur-sized telescope you can not even imagine the sight of thousands of stars packed tightly together in spherical balls. Though a six-inch will show all of them it will not resolve all of them. In some of the gobulars the stars are packed so tightly together in the central region that even with a large telescope this area looks like a bright-hazy, ghostly, grey-white area.

As with the open clusters the brighter and more prominent globulars are given M-numbers. Let's go over a number of the most interesting ones.

Object M-2

This is a seventh magnitude globular in the constellations of Aquarius. It is twelve minutes in diameter. Through a six-inch some of the stars around the edges can be resolved, but for the most part it looks like a bright, glowing mass. An eight-inch reflector will resolve stars about halfway toward the center and will show this object to be very bright. The stars will look very tiny when compared to the background stars in the field of view.

This is because this cluster is located outside our galaxy as are all globulars. Use about 80X to get the best view, but you may want to try about 100X with an eight-inch. A four-inch refractor at 50X will also show it very well and will probably resolve more of the stars than the six-inch reflector. The four-inch refractor however will not show as bright an image when compared to the reflector.

Object M-3

A beautiful cluster in Canes Venatici, there is a very bright central core in this one. It is of the seventh magnitude and is twenty minutes of arc in size. M-3 is round and through a six-inch the entire outer layers of the cluster can be resolved into countless numbers of stars. It gives something of a soft white appearance and the stars look very fine, as if they were crystals of salt sprinkled on a black surface. An eight or ten will resolve about 80 percent of the entire cluster and it will be a sight you will not soon forget. It is a beautifully fascinating cluster that is very easy to see in a small telescope. Since it is high up it is a simple object to observe.

Object M-4

A great globular located in Scorpio, M-4 is of the sixth magnitude and is twenty-three arc-minutes in size. It is one of the better globulars visible in the northern hemisphere and can be entirely resolved in an eight-inch reflector. A six-inch will resolve about 80 percent of it at about 50X; a three- or four-inch refractor will also show it very well. It is visible in binoculars as a hazy soft white patch. It can be found very easily since it is very close to the bright star Antaris.

Object M-5

This unique globular cluster lies in the constellation of Serpens and has an apparent magnitude of six. It is a moderate-to-large-sized globular cluster with a diameter of twenty minutes. This cluster is very beautiful and a six will resolve about 70 percent of it. There are quite a few stars in this one and the outer edges of the cluster will resolve into thousands of stars. An eight or a ten will resolve the cluster almost to the very core. Use 40–80X on this one.

Object M-9

This is a fairly small globular with a bright central core. It shines in the sky at magnitude eight and has an apparent size eight minutes of arc. A six-inch reflector will show it has a fuzzy ball of light, while an eight or ten will begin to resolve the outer edges. A large amateur telescope will be needed to resolve anything at all around the central area. In any case you will have to use at least 80X at minimum to start to see anything other than a blob of ghostly white light. M-9 lies in the constellation of Ophiuchus.

Object M-10

This globular cluster is also located in the constellation of Ophiuchus, but is much larger and somewhat brighter than M-9. M-10 has a size of twelve minutes and an apparent magnitude of seven. The core of this globular is slightly elongated as in M-5, but in this case it looks more pear-shaped. A six-inch will only show a fairly bright blob of light on good nights. Some of the outer edges may be resolved into stars. An eight will resolve the entire outer edges into countless stars and a ten or twelve will show the entire clusters as a myriad of stars with the exception of the central core area. To get the best view look at it with at least an eight-inch reflector or a four-inch refractor at about 50X.

Object M-13

The great cluster in Hercules. M-13 is a sight to behold; Fantastic even in a small telescope. This cluster has a diameter of about twenty-five minutes and is a bright magnitude of 5.5. It can be seen as a blob of hazy light through binoculars. A small 2.4- or 4-inch reflector will resolve the outer edges into hundreds of stars. A six-inch reflector will resolve the entire cluster except for the central core area. If you are lucky enought to use an eight-inch or larger 'scope you will see something that you will not soon forget. The entire cluster will resolve into thousands of stars. There is no way that I can describe the sight of M-13 through a medium or large 'scope, you will have to see it for yourself. Eight stars can be seen in front of the core area. This gives the cluster a three-dimensional effect that is breathtaking. You will also notice that the stars in this cluster appear larger than the others. This is because it is closer than most other clusters and is quite large. Even very low power will show it well. You might want to experiment with it at different magnifications.

Object M-15

A seventh-magnitude globular in Pegasus, M-15 is about thirteen minutes in size. If is about half the size of M-13, but it has a very bright central core area. Through a 12X finder it can be seen as a circular patch of greyish white light. A six-inch reflector will resolve only the very outer edges and at least a ten will be needed to get a really good look at it. Through a ten at 50X it will be very bright and about 70 percent of the entire cluster will resolve into thousands of stars that will look like fine salt grains poured on black velvet. A four-inch refractor will also give you a great view of this globular cluster.

Object M-19

Located in Ophiuchus, M-19 is very faint at magnitude eight and a tiny five minutes of arc in size. A six-inch reflector will show only a spherical blob of whitish light. Through an eight-inch some of the outer edges may be resolved into stars on a good night about 100X. A large amateur telescope twelve inches and above still will only show about 50 percent of this globular cluster. If you are using a refractor you might need a giant six incher to get a view of this one.

Object M-56

It looks about the same as M-19 but it is much fainter at magnitude nine. M-56 will only be seen well in at least a twelve-inch reflector. Even if you are using this larger 'scope you may only resolve the outer edges of the globular cluster. M-56 is an interesting object because it lies in a very dense star field. Through an eight-inch at 50X you can see this tiny blob of light surrounded by about a hundred hot blue-and-white stars and this adds considerable contrast.

Object M-62

Another small globular is M-62 in the constellation of Ophiuchus. This one is five minutes in size and has an apparent magnitude of seven. This one can be seen easily in a six-inch reflector, and all you may notice is how bright the center of the cluster is in comparison to the faint wisps around it which are unresolved stars. An eight will resolve some of these wisps into stars on a clear night. A ten-inch reflector shows it fairly well at 80X, as many of the star members start to become more visible and plainly seen.

Object M-92

This globular is also quite impressive. M-92 lies in the nothern part of the constellation of Hercules, shines with a magnitude of seven and is twelve minutes in diameter. This is a very compact globular with a bright core. One thing you may notice is that the outer edges contain fewer stars than most other globulars. This is because the cluster appears to be very densely packed. A six will show you just about all you would want to see on this one although an eight-inch reflector or a five-inch refractor will give a spectacular view of this fine object.

Object M-107

For the last of the globulars that I shall mention we once again turn to the constellation of Ophiuchus. M-107 is a faint magnitude nine and eight minutes in size. This object will look very faint and diffuse. You might even mistake it for some type of nebula instead of a star cluster. A six-inch will only show an irregular patch of faint hazy white light. An eight will show about the same, only brighter. With a twelve-inch at 100X you may start to resolve the outer edges, but only the very outer edges in the less dense star areas will resolve. This is a hard cluster and unless the night is very clear with no moon you may have trouble even finding it with a six-inch reflector. This will be also true if your area is plagued with the growing problem of light pollutions.

This ends the chapter on observing the stars. I must point out that the objects that were mentioned do not cover all the star clusters, open and globular, seen in the sky. By using a Messier chart and a star map you will find many more, especially in the constellation of Sagittarius. This constellation is somewhat low on the horizon for most of the United States and many of the objects may be lost in the natural atmospheric glow. However southern parts of the country do see this constellation high enough to observe the objects without too much difficulty. The higher in the sky above the horizon you observe a deep-sky object the clearer you will see it.

In the next chapter we shall cover observing another type of deep-sky object, the nebulae—glowing clouds of gas throughout the galaxy.

V

OBSERVING THE NEBULAE

Nebula are regions of dust and gas located in our galaxy. They appear through the telescope as fuzzy areas of light. The gas may be very dense and may look like a bright cloud, or it may look like small wisps of light white smoke.

The nebulae may come in any shape and they play an important part in the life and death cycles of the stars. There are three basic types of nebulae that can be seen in the night sky and each type having a different cause for its glowing. The three types are also different slightly in chemical composition.

Reflection Nebula

These are mostly faint nebula that shine by reflected light from a nearby star. If it were not for stars illuminating the nebula of this type it would barely be seen. The stars' light through the nebula becomes scattered, giving this type of nebula a bluish appearance. This type of reflection nebula is common near young stars as in the star cluster M-45.

Emission Nebula

These are glowing nebulae whose gases are excited to fluorescence by ultraviolet radiation from nearby stars. Radiation, usually from a very hot nearby star, hits the gas of the nebula, the atoms absorb the energy from the star then emit a glowing diffuse light. Emission nebulae can be divided into three types.

The Planetary Nebulae

This is a shell of gas ejected from a dying star. This shell is sometimes circular in appearance or disk-shaped—this is why they were first called planetary nebulae. The central star from which the gas came excites the nebulae to glow. Most of these types of emission nebulae are disk shaped, but several are not.

Nebula Wisps or Loops

These are just streaks of gas which arc into semicircles. They shine red from emission of radiation.

The Diffuse Nebulae

Normally very dense clouds of gas which may sometimes look like the thick cumulus clouds seen in the atmosphere of the Earth. In the center area of the diffuse nebulae are sometimes stars being born—it is these "baby" stars which excite the nebula to glow. Diffuse nebulae are the brightest nebulae in the sky and the easiest to observe and photograph. Diffuse nebula like the great nebula in Orion (the hunter) are called stellar incubators, the birthplace of the stars.

Dark Nebula

The so-called dark nebulae are nebulae that are black in appearance. They can only be seen if they are in contrast to a bright star field or a bright nebular region. They appear to be holes in space since the light from the stars does not penetrate the nebulae. Dark nebular regions can also be seen in some diffuse nebulae. Here the light from the exciting star is not reaching that portion of the nebula making it appear black.

To observe the nebulae one must be clear of city lights; a clear moonless night should be picked. You might also want to sue the nebula filter as described earlier in the section on amateur astronomer's equipment. Nebula filters will cut down light pollution considerably and transmit the nebular light frequencies making the nebulae brighter and more distinct.

The larger the telescope the better; however, the reflector will perform much better since the primary objectives are larger in comparision to amateur refractors. This will allow much greater lighter gathering ability which will show the nebulae brighter. A six-inch refractor is minimum, and even this will only show the much brighter ones. The fainter nebulae will only appear as blobs of diffuse light with a six-inch. An eight-inch reflector will show just about all the nebulae that are obtainable with an amateur-sized telescope. If you plan a serious visual or photographic study of these objects then you will need an eight-inch or a ten-inch.

In some cases the short focal lengths or the RFTs will work better, but in other cases a longer focal length will be needed. This will depend of course on the type of object that you plan to observe since the nebulae are of many different sizes.

I shall cover some of the more interesting nebulae within reach of an amateur astronomer's telescope. Most nebulae that we will be interested in will have an M-number; several will not. Nebulae are classified as deep-sky objects, but they are still located within our galaxy.

Object M-1

This is the famous Crab Nebula. This was the first object observed by Messier in 1758 and one of the most interesting nebulae for the scientist, but it is a great disappointment for the amateur astronomer. It is an oval-shaped cloud with an apparent angular size of four by six minutes and a magnitude of nine. This makes it very faint and you will need at least an eight-inch reflector to get a decent view of it. To see it well, a ten or twelve might be needed. This will depend upon the light pollution in your area.

What we see is an irregular green cloud. If you have a nebula filter by all means use it—you will need all the help you can get to observe it. I have seen it through a ten-inch reflector with the use of the nebula filter on a clear dark night, and I saw it quite well. I could clearly make out the structure and see the lighted filaments streaking out in all directions. It is called the crab because its shape was described by Messier as like a crab. I must confess for years I never saw any hint of this crab shape until I started using the nebula filter. Since I have been using the filter I do indeed see the shape described by Messier.

I live only thirty miles from New York City and the light pollution in my area on a scale from one to ten is about 8.5. It's ironic that Messier observed this object back in 1758 and saw it much brighter and clearer with a very small telescope. Of course back then there were no light pollution problems and skies were dark all over the country, but today much larger and better-quality equipment is needed to help combat the light pollution problem. I'm afraid that we are losing the battle; light pollution threatens the future of all visual and photographic astronomy.

Let's get back to the crab. The nebula itself is the sight of a supernova explosion that was observed in the east in the year 1054. The astronomers of China reported a "guest star" which appeared in the constellation of Taurus and was so bright that it was visible in the daytime. Astronomers today turn their telescopes toward this area and observe what is left of the star that blew off its outer shells of gas. That crab nebula is about 3700 light-years from the Earth and will appear green-white in an eight-inch at about 80X.

Object M-8

This beautiful nebula is located in the constellation of Sagittarius. M-8 shines at a magnitude of six and is a giant ninety by forty minutes in size. A six-inch f/4 will show it to be very distinct and bright. You will need the RFT-type reflector since this nebula is large in size. In order to appreciate the full beauty of it you must observe it all, and that means a richest field telescope and low power. M-8 is also called the Lagoon Nebula because you can plainly see the patches of dark nebulae scattered throughout the nebula. It is about five thousand light-years away and the actual size of M-8 is about 100 light-years across. If you are using an eight-inch f/4 or f/5 then you will be in for a treat. This is a fantastic bright object which will astound.

Once again if you live in an area of heavy light pollution, attach that old nebula filter. In any case, a nebula filter will give you a better view of it no matter where you live. M-8 is being illuminated and emitting the light absorbed by a number of hot blue-white stars which are close by. One of these stars is nine Sagittarii. For observers who live about forty degrees latitude M-8 will be somewhat low, and you may not see the entire nebula due to haze. Even here it is still a great sight.

Object M-17

This is also located in the constellation of Sagittarius. This is ofter referred to by astronomers as the Swan Nebula. It has a large apparent size of about forty-six by forty minutes and can be seen easily with a six-inch reflector. I suggest that you use the RFT-type reflector at low power for best results. M-17 has a very unusual form when viewed through an eight-inch reflector. It consists of two loops that form what appears to be the Greek letter omega. It is also ofter referred as the Omega Nebula because of this feature. M-17 is of the emission type and when photographed with color film it will appear red.

Object M-20

This is also of the emission type and is a diffuse, irregular, bright nebula. This remarkable nebula lies in the constellation of Sagittarius and has an apparent magnitude of about five. It is also large, as with most diffuse nebulae, at about thirty by thirty minutes (the entire area of the nebula is about equal to that of the full moon). M-20 is called the Trifid Nebula since it appears to be divided into three parts when viewed with a medium-sized telescope. Although a six-inch will show it, an eight-inch reflector will be needed to see the shape and the entire nebula. Use low power since this is a large object.

It is about 4.7 thousand light-years away and its diameter is about ten light-years. A nebula filter works well with this object and if you are using a small six-inch reflector you might need the filter to get a good look at this very beautiful object.

Object M-42

The greatest nebula in the northern hemisphere. M-42 is located below the third star in the belt of Orion. It is of the diffuse emission type and, at magnitude four, is visible to the naked eye as a small green-white patch of hazy light. It is very large at seventy by sixty minutes and a great sight even in binoculars. It is a very bright dense cloud and the sight of it through a medium-sized reflector may remind you of a large cumulus cloud on a sunny day.

Faint wisps of nebulosity can be see stretching quite a distance from the central point of the nebula. The actual main portion of the gas and dust cloud is expanding toward the Earth. This movement of direction is obvious since the main portion of the object is shifted in one direction. This is unlike other types of nebulae which are basically expanding out in all directions. Many dark nebulae can also be seen in this object. Near the center you may see several stars, these are newly forming stars or "protostars" which the nebula is giving birth to. M-42 is about 1,700 light-years away and is an easy object to observe and photograph. Close to M-42 you will see a fainter smaller nebula, this is M-43 and it is also of the emission type. A six-inch will not show it well but a ten will.

Object M-57

This is a planetary-type emission nebula located in the constellation of Lyra. M-57 is just about circular in shape and although it has an apparent magnitude of nine it appears bright.

Through a six-inch reflector it looks like a star out of focus at low power. To observe this object you will need at least 60X but 100X may work better with a larger telescope. M-57 is tiny and its size is only one minute across. Through an eight-inch reflector it becomes a most interesting object. It will appear somewhat oval in shape and a center hole can be seen. This gives the object a doughnut effect, as with most planetary nebulas of this type. The central star that illuminates the nebula is a hot ultraviolet star and has apparent magnitude of about thirteen. It can be seen with a small or medium telescope but may be photographed with an eight-inch with about an hour exposure.

The nebula filters will work very will with this one and with the filter it will appear twice as bright and much more detail can be seen. Using the filter with a six-inch reflector will give you about the same view as with an eight-inch without the filter. M-57 will appear blue-green, but some may only see an off-white with a little blue-green mixed through it.

Object M-27

This is another planetary type nebula although it is very different in shape to M-57. M-27 is often called the dumbbell nebula or the hourglass nebula. This is because the shape of the object is more that of an hourglass than anything else. M-27 is located in the constellation of Vulpecula and is the second largest planetary nebula in the sky. It has apparent magnitude of about eight and is four by eight minutes in size. A six-inch will show it having two large diffuse portions of nebula on each side with a wisp of nebula between them. An eight or ten will show the true hourglass shape and with this size telescope it will appear somewhat of a green-white in color. Using a nebula filter will make you wonder if you are observing the same object. Very faint filaments can be seen circling the nebula and now it will appear almost circular and very bright. It will look very dense with the filter at about 50X and if the night is very clear and dark you may see what appears to be dark nebula in the outer edges of the filaments. M-27 is one of the best objects in the sky since it also sits in a very heavily populated star area. This effect will add constrast, dimension, and beauty to the nebula.

Object M-76

Located in Perseus, this very faint nebula is of the eleventh magnitude and is about the size of the M-57 nebula in Lyra. If you have a six don't bother looking for it—even if you find it you will only see a faint little dot. An eight will show it a little brighter, but you most likely will need a large 'scope like a twelve or fourteen to see it as well as a six-inch would see the M-57 planetary. If you are using an eight-inch and you have a nebula filter use it. It may make the difference in seeing it at all.

Object M-97

This is a planetary type nebula located in Ursa Major. It is a very faint object but interesting. It will look like a circular puff of greenish white smoke in an eight-inch. Using the nebula filter with an eight-inch will darken the background sky and show the object much better. But unless you have a clear dark night and unless you are away from street lights you may have trouble seeing it with an eight. A six shows it as a very small wisp of gas just barely visible. A ten will show it brighter and clearer, a twelve-inch will show it even better. You will need at least an eight-inch to get a glimpse of it. Use about 50X and before you even attempt to look make sure that your eyes are adapted to the dark.

Planetary Nebula In Aquarius

Although this is the largest planetary nebula in the sky it is one of the faintest. It has an apparent size of about half that of the full moon. Through a six-inch it will appear as a smoke ring with a large central area. An eight might be needed to see it well; this will depend of course on your location. Use low power—40–60X with an f/5-8 reflector; a six-inch f/10 at about 50X may also give you a decent view.

Around a number of the brighter members of the Pleiades you may see faint wisps of a blue-white gas that surrounds them. This is a refection type nebula and with an RFT-type 'scope of eight inches or more it can be faintly seen. Using blue-sensitive photographic film with a blue filter it will show up quite well on a photographic plate with at least a twenty-minute exposure at f/4. If you are observing with the standard six-inch f/8 reflector you may see the nebula as a little fuzz around the stars in question.

The Rosette Nebula,

Like the two above, does not have an M-number, simply because it was not discovered by Messier. It does have an NGC number, however. This stands for New Galactic Catalog and is a modern attempt at listing the deep-sky objects. There are something like 7,000 NGC listings in comparison to Messier's 107 objects. It must be noted that the M-objects also have NGC numbers. The Rosette Nebula lies in the constellation of Monoceros and is about 2.5 thousand light years from Earth. It has a diameter of thirty-five light-years and an apparent size, as seen from the Earth, of about three full moons. It is quite bright in the sky and a good pair of binoculars will show it well. Through a six-inch reflector it will look like a ball of gas at about 50X. An eight-inch will show more structure; through this size telescope it may look like a fully bloomed rose. A central area where the material is less dense will be seen to enhance the rose-like appearance.

The Rosett Nebula is of the diffuse emission nebulae and is one of the more popular deep-sky objects for the amateur astronomer.

This concludes my coverage of selected nebulae to observe. The sky is full of all types of nebulae and once again they can not be all covered. I did however cover most of the brighter and more interesting ones. Next we will leave our galaxy and turn our 'scopes to other galaxies—the so-called island universes which are millions of light years away.

VI

OBSERVING THE GALAXIES

Galaxies are systems of hundreds of billions of stars, dust, and gas bound together by gravity. Our galaxy, which contains about 100 billion stars is but one galaxy in the universe. It is thought that there are about 200 billion galaxies in the visible universe, each with any where from 5 to 300 billion stars in each of them. It's quite a large universe! Astronomers when asked about the number of stars in the universe often reply, "A hundred billion stars in a hundred billion galaxies."

There are three basic types of galaxies; each one represents a different stage in galactic evolution.

Spiral Galaxies

These galaxies have a very bright core and are surrounded by spiral arms of dust, gas, and stars. The arms of a spiral galaxy may appear disk-shaped or they may be very wide and the galaxy may look like a child's pinwheel in space. These are usually interesting objects for the amateur since the spiral arms enhance the image. Our own galaxy, the Milky Way, is a spiral whose arms are closed and is disk-shaped in appearance. The bright band of fuzzy light that runs through the northeast to the south during the summer months is actually the central core or nucleus of our galaxy. This band of white light is called the

Milky Way and is really millions of unresolved stars, dust and gas.

Elliptical Galaxies

These galaxies do not have spiral arms; they appear elongated. They range in size from two thousand light-years across all the way to over ten times the size of our own galaxy which is close to 100,000 light-years across. Ellipticals are less interesting than spiral galaxies since they appear as fuzzy oval blobs of light.

Irregular Galaxies

Irregular galaxies have no shape, and they may look somewhat like a very tiny, faint nebula. Most of them seem to be disrupted by some type of explosions in the core area. They are interesting to observe if you have the right-sized telescope.

To observe galaxies you must pick a very clear dark night away from any street or area lights. You will be turning your telescope to objects which are millions of light-years away from our own galaxy.

One must remember that galaxies are millions of light-years away from us and that they are a very difficult type of deep-sky object to observe. A six-inch reflector will show only a few of them and these few are of the easy spiral type. All other galaxies will appear to be nothing more than faint blobs and patches of light when viewed through a small reflector. The reflector however does seem to be the better choice when looking at the galaxies. This is because you will need a large aperture in order to even begin to make out their structure.

You will need an eight-inch f/5-10 reflector. This is the minimum size telescope; a ten or twelve or larger is preferred. You can just about begin to make out the structure of about half the visible galaxies with an eight-inch reflector; the others will appear more or less as blobs of light with some shape to them.

Photographically, an eight-inch can resolve the structure of about 75 percent of all the galaxies within range of amateur-size equipment. The catch is that you must use special film with a fast system (about f/5) and a minimum of a one hour exposure.

Galaxies appeal only to the amateurs who are experienced enough to understand what they are really looking at. Once you realize that you are viewing the combined light of a hundred billion stars at a distance of millions of light-years will you appreciate it more. It is also interesting to note that since the galaxies are at great distances and light may take as long as ten million years to reach us, we are looking at the galaxy as it appeared ten million years in the past. Let's take a look at some of the better brighter galaxies in the northern hemisphere. We shall discuss them as they appear in an eight-inch reflector. All the galaxies will have M-numbers and are listed in the Messier catalog.

Object M-31

The brightest and largest galaxy in the sky. It is located in the constellation of Andromeda and is of the spiral type. It has an apparent magnitude of four and a size of 160 by 40 minutes. It is visible to the naked eye as a faint patch. An eight-inch will show a bright central core area with closed spiral arms. This object is so large that it may run out of the field of view in an eight-inch f/5 at 25X. Since it is so large you may want to use an RFT. Even a six-inch f/4 will show that one well. It is one of the only galaxies that can be seen well in a small telescope. With an eight-inch you will see dark dust lanes through the spiral arms. M-31 is photographically an easy object since with an f/5 system you can produce a fantastic picture with about a fifteen-minute exposure. It is very bright and one of the easiest galaxies in the sky to observe. M-31 is a close 2.2 million light-years away and has a diameter of about 180,000 light-years. It is the closest galaxy to us that is shaped similarly to our own galaxy.

Object M-32

This is a small elliptical galaxy that can be seen in the field of view of M-31. M-32 is a companion galaxy with an apparent magnitude of nine and a tiny size of two by three minutes. It overlaps one of the spiral arms of M-31 and is quite a hard object to see if the conditions are not right. Also in the same field of view you will see the galaxy which will be furthest away from M-31. This is NGC 205, another faint elliptical galaxy.

Object M-33

This is a spiral galaxy in Triangulum. It is very similar also to our own galaxy but the spiral arms seem to be more opened. Through an eight-inch reflector it appears as a large hazy patch with an irregular central core and dark gas lines. Around the core in a sort of loop, two of the spiral arms can clearly be seen. They will appear as faint wisps of glowing gas stretching out from the core area. Although M-33 is large at forty by sixty minutes it is very faint. In order to get a good look at it you will need a very clear, dark night with no moon or street lights lighting up the sky. If you live in northern Canada you will have no trouble at all seeing this galaxy, but if you live near a major city then the light pollution might wash out a great part of the spiral arms.

Object M-51

Beautiful spiral galaxy in Canes Venatici. This galaxy has very open spiral arms and is a classic example of the spiral-type galaxy. Through an eight-inch you will see a bright core with a fuzzy circular area around it. This is also part of the nucleus. Stretching out from the nucleus in a classic pinwheel fashion are the spiral arms. Although they will appear thin they are very bright and will seem to display a winding or turning effect as they circle the central core area. Stretching out from one of the arms is a faint wisp of gas with some dark dust lanes in it. This extension connects to a small satellite galaxy which has a very bright central core, NGC 5195. M-51 looks quite good at about 50X although for more resolution you may want to try about 90X. The galaxy measures six by twelve minutes in apparent size and although it is a faint magnitude nine seems very bright.

Objects M-58, 59, 60, 86, 87, 88, 89 and 90

These are located in the constellation of Virgo. All of them except for M-58 and M-90 are elliptical galaxies; these two are spiral. They all appear basically the same and unless you have a fourteen-inch telescope you may not be able to tell the spirals here from the ellipticals. They all have an average size of about four by five minutes and a magnitude of ten. They will appear as fuzzy objects with bright central core areas.

Object M-61

This is a spiral galaxy in the Virgo area. It is of the tenth magnitude and is six by six minutes in size. Through an eight-inch reflector at about 80X the central core area looks quite bright, as with most spiral galaxies, and you will see it surrounded on both sides by what may appear to be very faint glowing extensions. These areas are the spiral arms and an eight will just barely show them. A twelve is needed to show them well.

Objects M-65 and M-66

These two galaxies lie very close together and with an eight-inch at low power you will be able to fit them easily in the same field of view. Both are spirals located in the constellation of Leo and both have an apparent magnitude of about ten. They are both two by eight minutes in size and although they appear very thin they will appear very long. They are both interesting in an eight-inch reflector and look like glowing lines of gas in space. I have seen these two quite well in a four-inch refractor at 50X, but the eight will show them better, and a twelve-inch reflector will give you a great look at them.

Object M-74

This spiral galaxy in Pisces is moderate in size at eight by eight minutes, however it will appear faint since its total magnitude spread out across this distance is ten. Through an eight it will appear as a circular patch of faint glowing light. All you will be able to see is the central core. An eight will not show you the spiral arms. Use about 50X to look for it; if you have setting circles look up the coordinates in the Messier catalog and turn your 'scope toward these circle settings. Always use your circles when looking for faint galaxies; it may make the difference in finding it and not finding it at all.

This galaxy is very faint so do not expect to see a brightly glowing cloud and spiral structure. You will only see the bright nucleus and at best it will only appear as a faintly glowing blob of light.

Object M-77

A very small spiral galaxy in Cetus, M-77 measures only two by two minutes in apparent size. Since it is so small looking it will be fairly bright at magnitude ten, but once again all you will see is a bright core area. Through an eight you might notice some slight glowing haze around the edges. These are parts of the halo around the galactic core. They are not the spiral arms; the spiral arms are not visible in amateur-sized telescopes.

Object M-81

A truly remarkable spiral galaxy in Ursa Major. M-81 has an apparent size of sixteen by twenty minutes and a magnitude of seven. Through an eight-inch the nucleus of this galaxy will be very bright and the halo of dust lanes around the galactic core will be clearly visible. An eight will also show the spiral arms very well and will clearly show how they extend out from the core area of the galaxy. M-81 is about six million light years from the Earth and seems to be half face on toward the Earth.

Object M-82

M-82 lies not far from M-81 and a wide-angle RFT at low power should show them in the same field of view. M-82 is an irregular galaxy and as a matter of fact it is the only irregular-type galaxy in the Messier listing. M-82 is two by seven minutes and is of the ninth magnitude. It is horizontally fixed as seen from the Earth and will appear as a bright glowing green, blue, or white cloud, very thin but long through an eight-inch reflector. This is one of my favorite galaxies to observe since near the central area there seems to be some violent activity taking place. Here streaks of dark matter can be seen and it appears as if the galaxy is slowly being ripped apart. M-82, unlike the spirals, seems to have equal brightness throughout its length. Use about 50X to observe it. If you have an RFT try looking at it under low power for a wide angle view of both M-81 and M-82.

Object M-94

A spiral galaxy located in Canes Venatici, M-94 is five by four minutes in size and has an apparent magnitude of nine. As seen from the Earth this spiral is face on, but through an eight-inch will look like a circular wisp of glowing white light with a bright starlike core. The spiral arms are closed in this one so through the eight they will not resolve but appear like a faint circular or elliptical cloud around the core.

Objects M-95 and M-96

These are two spiral galaxies which lie in Leo, both of the tenth magnitude and very faint. M-95 is the smaller and brighter of the two and appears as a circular ball of gas and it may remind you of an unresolved faint globular cluster. M-96 at seven by three minutes is longer since it is edge on towards the Earth. It will be much fainter, however, and you will need a very clear dark night to see it.

Object M-101

With a diameter of twenty-two minutes this spiral ranks as one of the largest in the sky, but at magnitude eight spread over this distance it is very faint. It is located in Ursa Major and the spiral arms are very open and easily visible once you see it. That's the trick—seeing it. This object is so faint that unless you are deep in the country or least a hundred miles from any major city your chances of geting a good look at it are not good at all. The nucleus is small but much brighter than the entire galaxy.

Object M-104

An unusual spiral galaxy in Virgo, M-104 has an apparent size of seven by two minutes and a magnitude of eight. It is an edge-on spiral with very diffuse arms. Through the center extending through almost the entire length of the galaxy a very dense dust lane can be seen. This dark gas gives M-104 the appearance of a Mexican sombrero; it is often called the Sombrero Galaxy. Through an eight the galaxy can be seen very well as an object with a thick center and two fainter extensions of glowing gas stretching out in a linear fashion on both sides. An eight-inch will show the dark dust lane but only near the central area of the galaxy. This is because the galaxy is much brighter here and the dark lane will stand out better because of the contrasting effect of black on white.

I suggest using about 100X to observe it, and needless to say a clear dark night will show the galaxy at its best.

Object M-105

A tiny elliptical galaxy in Leo, M-105 is only two minutes in apparent diameter and of the tenth magnitude in brightness. Through an eight-inch reflector it will appear like nothing more than a fuzzy star with a bright core area. It is a hard galaxy to observe, but it is a good test for your abilities to recognize a very faint galaxy and pick it out from the background stars. Use about 50X since you will need all the light gathering power of your scope to pick it out. I have even seen this object with a ten-inch reflector on a night with perfect conditions and it really showed no more than the eight-inch did—it just seemed a little brighter.

Object M-106

A fairly good spiral galaxy located in Canes Venatici, M-106 is a large twenty by six minutes but at this size it is fairly faint at magnitude nine. It is somewhat edge on towards the Earth but seems to be more on a forty-degree angle. The central core area can be seen as a bright flowing mass, but the arms on this spiral galaxy are pretty hard to make out with an eight-inch reflector.

Objects M-108 and M-109

Two tenth-magnitude spirals in Ursa Major, both about the same size, both at different angles toward the Earth. M-108 measures eight by two minutes. Through an eight-inch reflector it appears as a thin glowing faint line. M-109 is face on and has a diameter of eight minutes. It will appear through an eight-inch as a faint circular haze, somewhat brighter in the central core. These two galaxies are a test for a novice amateur astronomer to find.

VII

BASIC AMATEUR ASTROPHOTOGRAPHY

Astrophotography, especially with long time exposures through the telescope, is not easy. It is the purpose of this chapter to go over some of the steps to begin astrophotography. Many times while photographing through the telescope you may only look at the object once or twice in an hour's time. Ninety-eight percent of this time will be spent guiding on a star close to the object you are photographing. Although you may spend hours of time making exposures, on an average of about 60 percent of your shots will come out. This is due to a great number of factors, from stray light to vibrations causing a blurring of the object. Below is listed equipment that you will need.

The Camera

A 35mm SLR with a removable lens is the best choice. When we shoot through the telescope we unscrew the lens and hook up the camera to the 'scope with a special adapter. These adapters can be bought through a number of scientific astronomical supply houses and camera stores. They carry many types for different cameras and 'scopes.

Cable Release

This screws right into the shutter button and is used to lock open the shutter for a time exposure. It also helps to knock out vibration since you will not have to touch the shutter button and camera at all to make an exposure. The cable release is fairly cheap and can be purchased at any camera store.

Tripod

To make constellation or star trail photos you will need a tripod to screw your camera into for support. To shoot the stars of various constellations you will not need your telescope, just your camera mounted on the tripod with a standard 50mm lens.

Film

There are many types of film on the market today. The standard films such as the Kodak series are quite good for a number of projects in astrophotography. The Ektachrome series is the best bet for the amateur if he is shooting with standard color film. I must mention that for astrophotography always use color slide film or black-and-white print film. Never use color print film since most commercial processors will print it incorrectly, giving washed-out images.

Ektachrome comes in a number of speeds or ASA numbers. The higher the ASA number the more senstive or the "faster" the film. The faster the film, however the more grainy it is. This means that with the faster films you will lose some fine detail in the objects that you are shooting. This does not really effect stars and deep sky objects but it will affect the clarity of your shots of the moon and sun and planets. Since deep sky objects are faint you need a very fast film; the planets are bright so you don't need a fast film. A slow film or fine-grained film should be used on the moon, sun, and planets. The Ektachrome film comes in three basic speeds. Ektachrome 64, which is a very fine grained but slow film, will perform well in the photography of the sun, moon, and planets. Ektachrome 200, which is fast but not too grainy, can be used on the planets and a number of bright deep-sky objects. It is also good for shooting stars and constellations. Ektachrome 400 and is very fast and should be used for the photography of the fainter objects.

Black-and-white films are also very useful in astronomy so don't be put off by them just because you want to see your picture in color. Since, here, you are only dealing with tones of grey and black you have more control over the production of your picture. Color film may at times give you off-color results due to light pollution. Light pollution can cause a greenish sky in your color pictures. Black-and-white will not show this and you get nice bright white star or nebular images on a black background. Also, with black and white film, you can use a red filter to knock out most of the skyglow. This will produce a brighter image, a darker background, and a better all-around photograph.

Standard Kodak films can be bought at any camera store and like the color films they come in all speeds from a fine-grained 32 to a very fast grainy 400. What we have covered is standard films, and although it can be used for photography of the sun, moon, planets, and a few globular clusters and nebulae, it falls short when you have to make an exposure of over ten to fifteen minutes. The films listed above will record the object in the same area of the emulsion for only this time, then stop recording. For example, a fifteen-minute exposure will look the same on the standard films as an hour exposure. After fifteen minutes the film just stops recording the image.

There is a solution, however. If you are planning long exposures of faint objects then you will need a special group of films offered to the amateur astronomer. These are called spectroscopic films and can be bought through a number of mail-order scientific companies. These films are all black and white, but they are sensitive to different parts of the spectrum. They all have an ASA rating of from 400 to 600 and can be used in exposures of an hour or better. There are three basic types, each with a different function in astrophotography.

103-AF is an all-purpose astro film sensitive to just about the entire visible spectrum. (This is called a "pan" film.) It works well for globular clusters, galaxies, and open star clusters, and will also perform well with some nebulae.

103-AE is a red-sensitive ("ortho") film and works well with most galaxies. It also works well with nebulae, especially diffuse emission type. It will also be a good choice for globular clusters. Using this film in conjunction with a red filter will help transmit the red end of the spectrum to which the film is sensitive. This will also make your image brighter and knock out some of the light pollution. For photography of emission and planetary nebulae, this film with a nebula filter is a good combination.

103-AO is a blue-sensitive film used for photography of reflection nebulae, some open clusters, and galaxies. It will work well with some objects with a light blue filter.

The films listed above are manufactured by Kodak and they are used by amateur astronomers for deep-sky photography. They will not work well at all for the planets, sun, or moon because they are grainy, and although the contrast is very good the resolution is poor. Another drawback is that this film must be kept under refrigeration when not in use. And this film is expensive when compared to the standard films.

For deep-sky photography one may start out with the standard films to get experience, but any serious work must be done on the above films. There is a way however to use the color films for long exposure and that is with the use of a special cold camera. This camera has a special chamber for dry ice which chills the emulsion of the film to a sub-zero temperature. This allows the use of standard films with very long exposures. The grain will be very fine, but due to its chemical nature the film will still be fast and very sensitive to light. With this procedure an amateur astronomer with a moderate-sized 'scope will produce photographs of deep-sky objects that will rival professional observatory photographs. (It is not the purpose of this chapter to cover cold cameras. Cold-camera photography is not basic astrophotography and if you want to know more I suggest that you contact one of the manufactors listed who carry this device.)

For planetary photography there is also a special type of film for the amateur astrophotographer. This film is called SO-115 and although it has an ASA of about 160 it has the grain of a film with an ASA of 20 or less. I have used this film and it will provide high-resolution photographs of the sun, moon and naked-eye planets. It is also manufactured by Kodak and will also provide very high contrast. It is also rather expensive, but unlike the 103-series films it is not used for deep-sky photography but only planetary, lunar, and solar photography.

Telescopes

Just about any telescope can be used for astrophotography, but your system must have an equatorial mount and must have a clock-driven RA axis. If you are planning deep-sky photography, that is an exposure of ten minutes or more in length, it would be worthwhile to have a variable speed control and a manual or motorized declination adjustment. (These are also called declination slow motion controls and are used to make minute adjustments in the declination.)

If you plan to photograph the planets, in order to achieve sharp highly resolved images you should have a focal length of at least sixty inches. If you plan deep-sky photography then you will need a fast (f/5 or better) photographic system.

Photographing Star Trails

This is very easy to do. Here's how to do it and what you will need:

1.) A 35mm camera with a standard 50mm lens at about f/1.4–4.
2.) Fast color or black and white film—ASA 200 or 400 will do nicely.
3.) A cable release.
4.) A sturdy camera tripod.
5.) A flat-black piece of cardboard about four inches square to help control vibration while the shutter is open.

Point the camera on the tripod to the section of the sky that you wish to make a trail photo of. Look through the SLR screen of your camera, frame the area, and lock the tripod so that it is stable. Attach your shutter release cable. Take the cardboard square and place it in front of the camera lens being careful not to touch any part of the camera. While holding the cardboard square here put the shutter speed on time exposure (B) then press the button on the cable release. All the time remember you are holding the cardboard in front of the lens. Lock the button on the cable release so that the shutter will stay locked open, count five seconds then move the cardboard away from the front of the lens area. By holding the cardboard square in front of the lens for this amount of time it will allow any vibration that was caused by the shutter opening to stop. Shutter vibration can cause a slight blurring and loss of clarity in your picture.

Expose a number of frames with different exposure times. It takes about fifteen minutes to get a good trail. If you live in a heavily light-polluted area, unless you are using black-and-white film with a red filter, your exposure will be restricted to about ten minutes.

Photographing the Constellations

Shooting the constellations is also fairly easy. There are two types of exposures. The first is a quick exposure of about thirty seconds. This will show the main stars of the constellation and they make very nice art and teaching photographs, especially on color slides.

The second is an exposure of five minutes or more—a guided exposure where the camera is mounted piggyback on the bottom or top of the tube of a telescope that is equatorially mounted and clock driven. If you are planning an exposure of more than one minute it will have to be guided and clock driven, since any exposures longer than forty-five seconds or so will show the stars as tiny lines rather than points of light.

Unguided exposures are easy—the directions and equipment needed are the same as for photographing star trails. Exposures are thirty seconds to one minute.

Guided exposures will show a great wealth of stars and with longer exposures you will see thousands of stars in your finished photo. In addition to the equipment already listed you will need an equatorial-mount telescope with clock drive and a means of attaching the camera to the telescope tube. As for film, color slides produce interesting shots but sky glow may limit your exposure if you live close to a near by city. In this case I suggest using red sensitive 103 ae film with a red filter.

Attach the camera onto the telescope tube or mount in any way it can be done. Make sure, however, the camera is on the RA axis if you hook it onto the mount. The benefit of attaching it to a telescope tube is that once you expose the image you can track it in the eyepiece of the telescope to make sure you are guiding well enough. You may need to if you have a poor polar alignment. The camera may be mounted on the telescope tube by rope, straps, or you can buy a special camera mount which allow this type of photography.

Point the telescope and camera to the section of the sky you wish to photograph. Look into the SLR screen of the telescope to make sure the camera is exactly where you want it. Look into your telescope and find a good star to keep in the field. Use about 150X. Hold the cardboard square in front of the camera lens, click the shutter using the cable release, lock the shutter open. Wait five seconds then remove the cardboard from the front of the camera lens. Look into your telescope and keep the star that you selected in the field of view. Expose anywhere from five to twenty minutes. Experiment as you frame your shots. Beware of sudden lights like a neighbor's porch light, aircraft, car lights, and so on. They will fog out your photograph.

Basic Solar Photography

Photographing sunspots and other types of phenomena on the sun is very worthwhile and makes impressive pictures. You will need:

1. A 35mm SLR camera with no lens. Here you must remove the lens since the telescope will act as the camera lens. You must use a camera that has the interchangeable lens feature, and an adapter to mount it to the telescope.

2. Cable release
3. Fine grained film, slide film ASA 64, B&W SO-115 or ASA 32 or 125.
4. Solar Skreen by Tuthill, or a similar device.
5. Yellow or green filter for color astrophotography.
6. Telescope: minimum three-inch refractor. (A reflector of six inches minimum may also be used but the refractor is preferred).

Although it is a good idea to have your telescope on a clock driven equatorial mount for high power photography, you will not need one at low power. Using the camera adapter hooked into your camera, slide the tube end of the camera adapter into the draw tube of the rack and pinion (focuser) of the telescope. We will be using prime-focus photography which means all we are going to shoot with is the telescope's objective which will focus an image of the sun into the camera without an eyepiece. This will give us low power, but enough to get some good high-resolution shots of the solar disk. Each telescope will have a different magnification at prime focus. This depends on the telescope's focal length. To find out roughly what magnification your 'scope will give, simply divide your focal length in millimeters by forty-five. For example, a three-inch f/15 refractor has a focal length of 45 inches or 1125mm (there are 25 millimeters per inch). Dividing 1125 by 45 gives us a magnification in our example telescope of 25X.

Attach the Solar Skreen to the top of the tube making sure it is on well and that there are no breaks in the material. Point the telescope toward the sun and look into the SLR screen of the camera. When you are near the sun the area will become brighter—just head toward the bright area. Soon you will see the solar disk, which with the Solar Skreen will appear slightly blue. Focus well, take your time. You should be able to see the sunspots even through the diffuse ground-glass SLR screen of the camera.

To get the proper exposure you may use the camera's built-in light meter; the right exposure may be found by changing the shutter speeds until your meter is in the proper exposure area. With ASA 64 film your exposure on a bright sunny day may be something like 1/250 of a second. However be sure to "bracket" your exposures with a number of different exposures to be sure. Don't settle for three or four frames, since atmospheric refractions and wind turbulence can wreck your pictures. At least ten frames will insure that at least several of them were taken during moments of atmospheric stability.

For black-and-white film no filteration should be used. For color film use a light yellow filter. This will give you a yellow colored image and will add contrast to the sun spots and bright areas better. The filter can be placed between the camera and the end of the camera's adapter by screwing it into a place provided for such a purpose. It may also be mounted in place in this location by using an empty eyepiece barrel which the filter will also screw into.

Photographing The Moon

Lunar photographs are quite common among amateur astronomers, but high-resolution shots of excellent quality are rare. You can actually map the lunar surface down to surface detail of less than one mile with a moderate-sized amateur telescope. The lunar terrain is also an impressive sight in a color slide. The equipment is the same as for solar photography, except for the film (somewhat faster films—ASA 32–200—are used). A 9mm eyepiece is used for high-power photography. For low power photography you needn't use the eyepiece. Here you will get the entire moon in the frame. The moon will vary in brightness according to the phase, but in most cases it can be metered with your camera's built-in light meter. Make sure that you shoot at several different exposures with several frames each. Attach the camera to the 'scope using the camera adapter, meter your exposure, and release the shutter with your cable release. For higher-power photos—close-ups of a specific area on the moon—magnification may range from anywhere from 150X-300X. I like to use about 194X with an eight-inch or ten-inch. The only difference from low-power photography is that you will be changing the exposure time and adding an eyepiece. Load your 9mm eyepiece (a 6mm, 10mm, or 12mm will also work well) into the eyepiece holder of the camera adapter, and focus on the moon, making sure that you have it in the best possible focus. Make sure also that you are near the area that you wish to photograph. Hold the cardboard in front of the telescope tube making sure not to touch it. With your other hand press down the button of the cable release. Count five seconds then carefully pull the cardboard away and make your exposure. Your exposure will vary in relation to the film speed that you use. With an eight-inch reflector at 194X with ASA 64 film on a first-quarter moon, the exposure is around three seconds. You have to shoot a number of frames and then evaluate your photographs and see what exposure seem to be the best. Needless to say, while you are taking the photographs you must take accurate notes of the exposure times frame by frame.

Photographing the Planets

The planets that we will photograph will include Mercury, Venus, Mars, Jupiter, and Saturn. The others will show only a very small disk and are not worth photographing for display purposes. Fairly high powers are needed for these photos (200X–500X) so large telescopes are preferable to small ones for sharp images. A 4–9mm eyepiece is also needed. Medium-speed films are sufficient.

Looking through the SLR screen of your camera focus on the planet very carefully. If you have an interchangable screen in your camera (as some of the more advanced cameras do) it would be good to put in a clear glass screen rather than the diffuse screens that are often supplied. At this high magnification it will be quite hard to get a proper focus on the planet's disk. Observe the image for several minutes to insure that your driving system is working properly. Place the cardboard in front of the telescope objective as before and your camera shutter speed on B (time exposure). Click the shutter open and hold the cardboard in front making sure you do not touch the tube of the telescope. Hold there for about ten seconds to made sure all vibration has died out. Carefully move the cardboard away and make your exposure, move the cardboard back in front of the tube, and then click the shutter closed. Your exposure time will vary anywhere from one to five seconds. This will depend on the speed of the film, the magnification used, and the planet you are photographing.

Remember you are shooting at very high powers and the slightest vibration can ruin your picture. Heat trails from passing jets can also cause turbulence to blur your photograph if they are close by. Once that shutter is open do not move at all, and be very careful. This is why if you are planning high-power planetary photography you better be prepared to shoot an entire roll of film.

Of twenty exposures, anywhere from six to eight may come out all right with only two or three excellent frames. The thing here is to shoot for detail and the best film seems to be SO-115 fine-grained film by Kodak. It will give you the speed of a fast film, but the grain of a slow film. A quick shutter speed is desired to minimize any unpredictable effects that could ruin your photograph.

You may also want to try using filters with the color and black-and-white films, but the contrast in the SO-115 is high enough that you will not need any filtration to increase the contrast. You will have to experiment with exposures and films to see what will work best for you in your area.

Photographing Deep-Sky Objects

Perhaps one of the most difficult aspects of astrophotography is the photographing of the star clusters, nebulas, and galaxies. Fast film (such as the Kodak 103 series) and an eight-inch reflector or its equivalent are needed. A good clock drive is also necessary. The gear size should be at least three-quarters the size of the objective mirror. The telescope system should be no slower than f/6; f/4 is better. Variable speed controls also help to keep the subject centered.

Your exposures must be guided through an illuminated recticle eyepiece. This is basically two lighted cross hairs forming a cross in the center of the field of view. Using the fast and slow controls of the variable speed controls you keep the star that you are guiding on in the center of the cross hairs. Your main scope will be on the deep-sky object that you are photographing while your guide scope will be off axis, trained on a bright closeby star to insure no streaking in the photograph. Different deep sky objects require different exposures, for example, (with an f/5 'scope, exposures for open cluster range from five to fifteen minutes; globular clusters—minimum fifteen minutes; nebulae (all types)—minimum thirty minutes; galaxies (all types)—minimum sixty minutes (although M-31 can be photographed well in as little as fifteen minutes).

For further information on deep-sky photography consult books on advanced astrophotography.

If no photo credit is given then the photo was by the author.

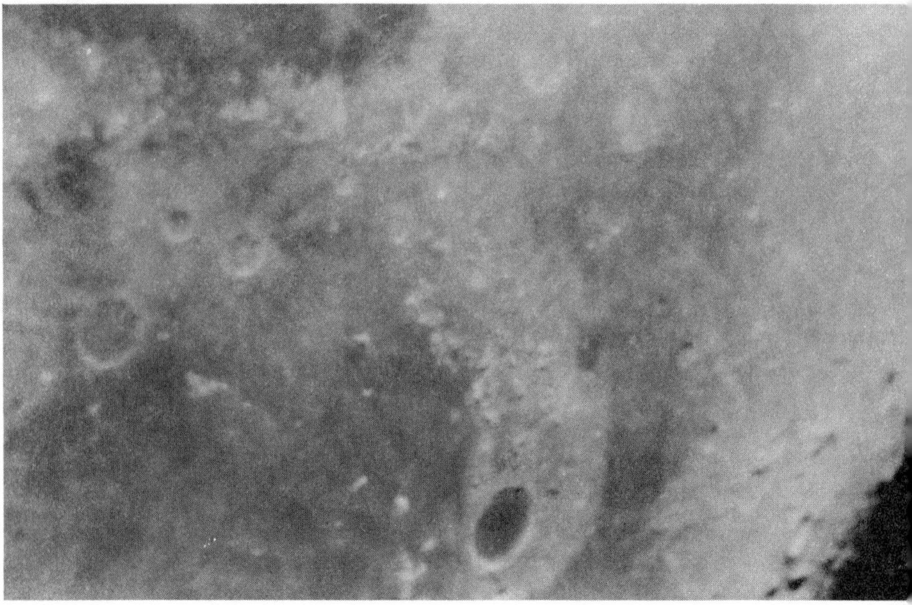

Plate 1. Area of the moon showing the crater Plato. A ten-inch reflector was used at 190X. The film was Plus-X by Kodak. Exposure time, 2 seconds.

Plate 2. The full moon. Notice the mare or sea areas and the bright crater rays. Shot with an eight-inch Schmidt-Cassegrain at prime focus at f/10 (about 50X). Tri-X pan by Kodak was used for 1/125 seconds.

Plate 3. The Southern Lunar Highlands. Notice the very rough crater areas. Taken with a ten-inch Newtonian reflector 190X with Plus-X by Kodak. Exposure time, 2 seconds.

Plate 4. The Mare Humorum connects to the large crater Gassendi. This outstanding photograph was taken with a ten-inch Newtonian reflector at 190X. Pan-X film was used in this 3-second exposure. *Photo by Cathy Salerno*

Plate 5. The lunar Altai Mountains seen as a bright streak. The three craters are, from right to left, Theophillus, Cyrillus, and Catharina. The mountains mentioned above are just above this area. Taken with a ten-inch Newtonian reflector at 190X using Plus-X film by Kodak with a 2-second exposure. *Photo by Cathy Salerno*

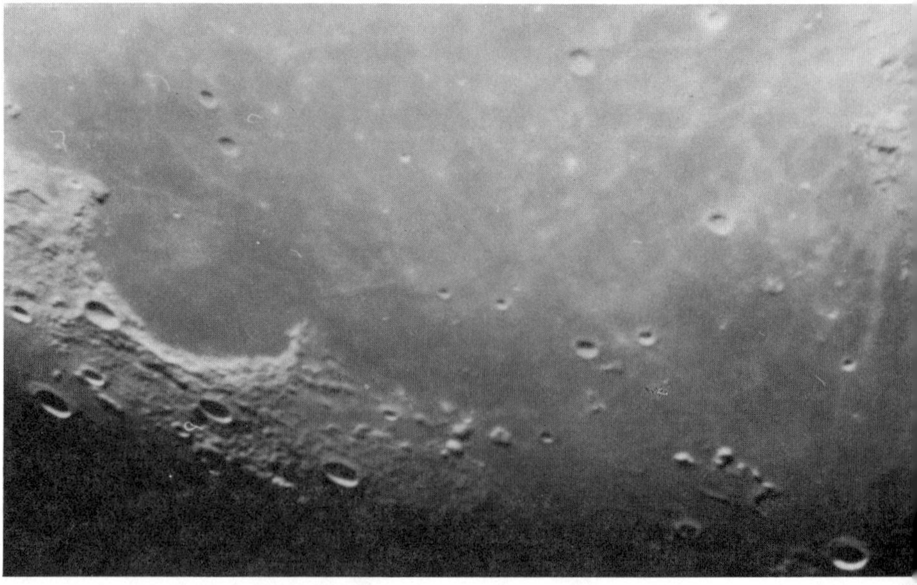

Plate 6. The northern area of the Mare Imbrium. Taken with a ten-inch Newtonian reflector 190X with Plus-X film for 3 seconds. *Photo by Cathy Salerno and the author*

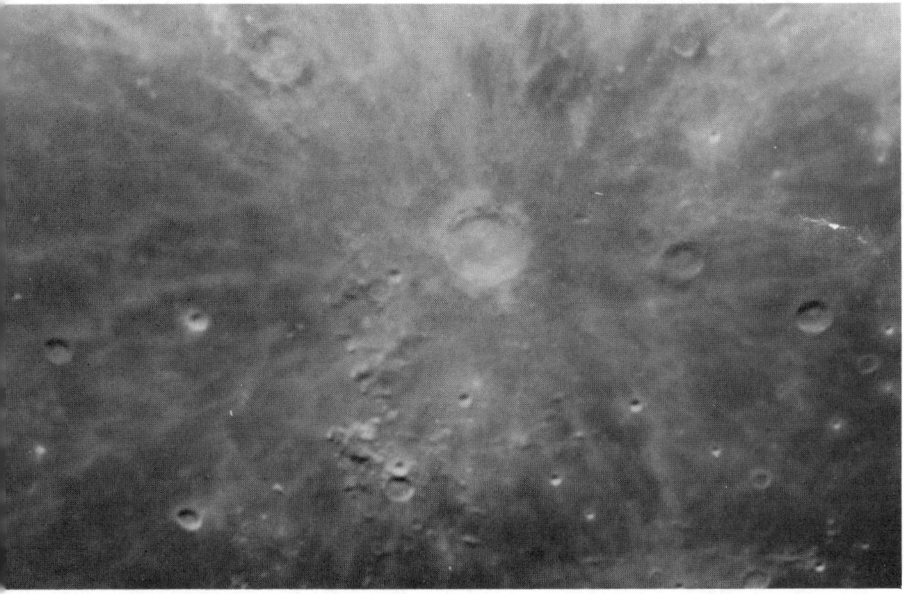

Plate 7. The bright-rayed crater Copernicus. A ten-inch reflector of the Newtonian type captured this outstanding lunar feature using Plus-X film at 190X for 2 seconds. *Photo by Cathy Salerno*

Plate 8. The sun showing sunspots. Photographed with a three-inch f/15 refractor at prime focus (25X) for 1/125 second on Kodak Pan-X film. Notice all the large spots—this photograph was taken during a sunspot maximum time.

Plate 9. The crescent phase of Venus. Venus as it approaches inferior conjunction will become a thin crescent. This was taken with an eight-inch Newtonian reflector at 234X for three seconds using high-contrast copy film by Kodak.

Plate 10. Mars during an unfavorable opposition. Notice how small the disk is even at 350X. During a favorable opposition the disk however will be over twice this size. Detail is still however visible and the center fingerlike area is the Syrtis Major. Photo taken with an eight-inch Schmidt Cassegrain using SO-115 film.

Plate 11a. Jupiter showing a large storm in the upper middle section.

Plate 11b. Jupiter showing a large eruption-type phenomenon moments later in the lower center area. Notice the position of the storm in both photographs indicating that the time between frames was not very long (about fifteen minutes). Notice the cloud band detail on the planet. This photograph was taken with an eight-inch Schmidt-Cassegrain at 300X using SO-115 film for four seconds.

Plate 12. Globular cluster M-13. This fine large cluster is located outside the galaxy about 25,000 light years away. This photograph was taken with an eight-inch Schmidt-Cassegrain at f/5 (about 25X). Exposure time was twenty minutes on 103 af film.

Plate 13. Globular cluster M-5. This fine cluster is further away than M-13 but is a great sight. This photograph was taken also with an eight-inch Schmidt-Cassegrain at f/5 (25X) for fifteen minutes on 103 af film.

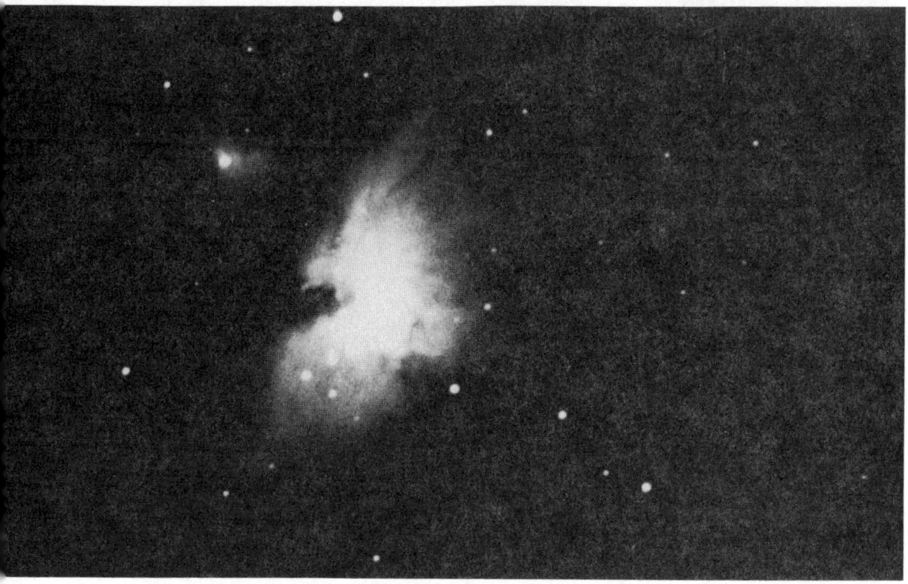

Plate 14. The Orion Nebula. Photographed with an eight-inch Schmidt-Cassegrain at f/10 (about 50X) for 40 minutes with 103 ae film and a nebula filter.

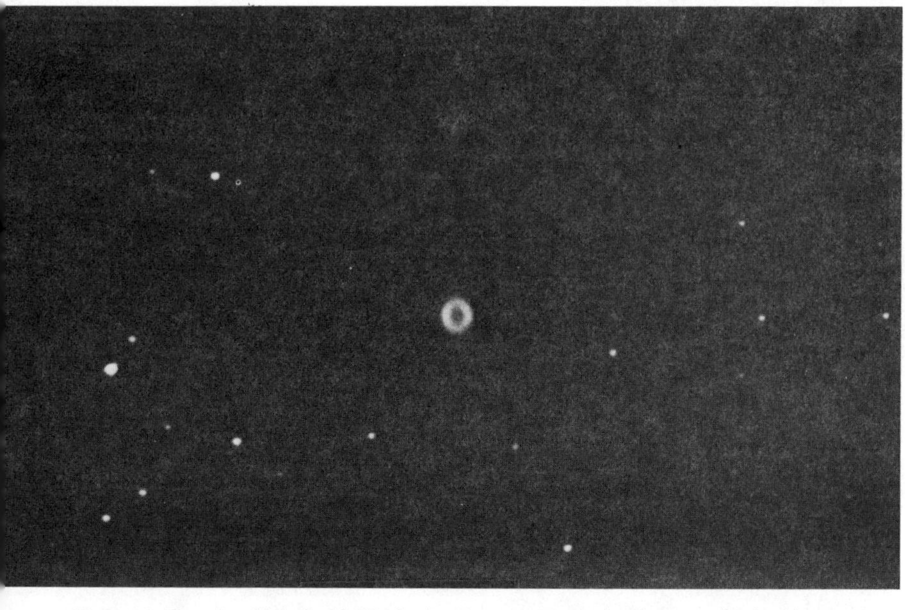

Plate 15. The ring Nebula M-57. A planetary-type nebula photographed with an eight-inch Schmidt-Cassegrain at f/10 with 103 af film for 45 minutes with a nebula filter.

Plate 16. Spiral Galaxy M-51. A loose spiral and a beautiful sight. Taken with an eight-inch Schmidt-Cassegrain at f/5 (about 25X) for one hour at 103 ae film.

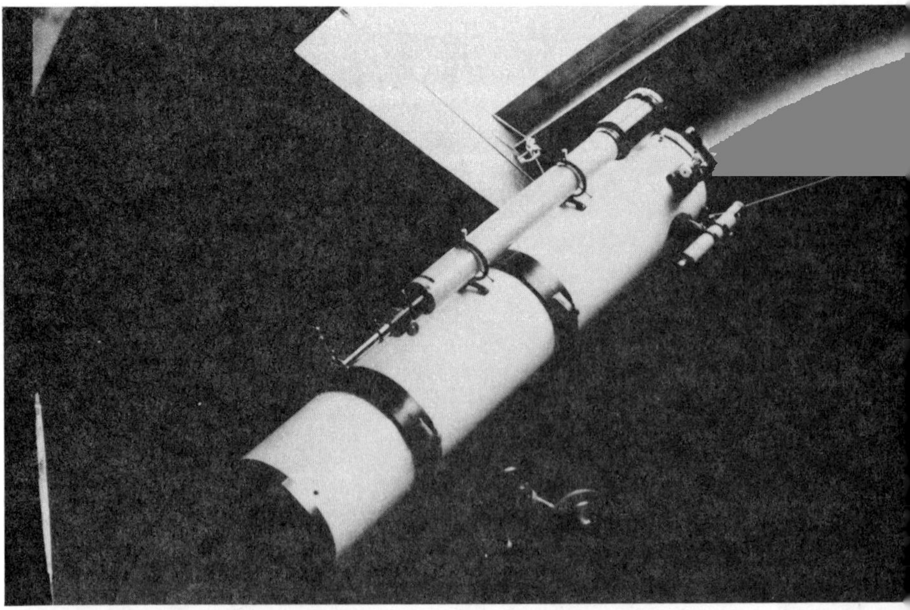

Plate 17. The ten-inch Newtonian reflector on a pier and German Equatorial mount in the Milbank observatory in Greenwich, Connecticut.

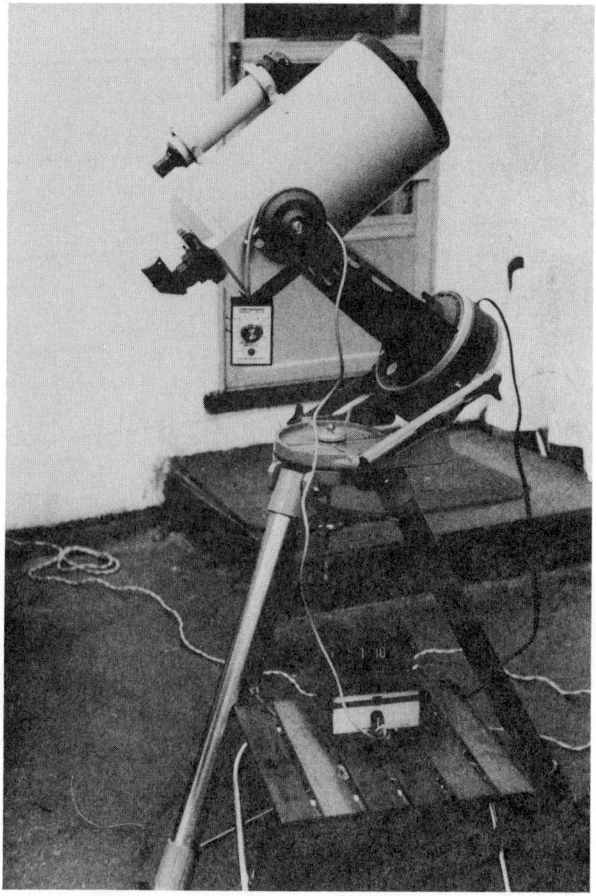

Plate 18. A compound telescope (Schmidt-Cassegrain type) on portable tripod—(an eight-inch telescope). Also visible are variable-speed controls; unit is mounted on an equatorial fork mount.

Plate 19. Rick Brias's four-inch Unitron f/15 refractor on equatorial tripod.

Plate 20. Jupiter and four moons. They are from left to right, Io, Europa, Ganymede, and Callisto.

Plate 21. Skylab as it passed over the Greenwich Connecticut area in June of 1979.

VIII

OBSERVING HINTS

In this chapter I will cover some of the basic problems you may encounter while using your telescope. Remember Murphy's Law, "If anything can go wrong it will." It is true with astronomy some nights—everything may go perfect and you may only experience some minor problems; other nights nothing seems to go right at all. This is especially true with astrophotography where your sucess is at best about fifty-fifty.

Always select a good night without possibility of rain, and always monitor the weather closely by listening to the reports on your TV or radio. You may want to purchase one of those special weather radios which give a twenty-four-hour forecast.

It is a good idea to set up your equipment while it is still light outside. This will avoid any accidents in the dark which you might regret later. Let your telescope adapt to the cooling night air. This is true especially in the winter. When the temperature starts to drop in the nighttime the telescope will start to cool. If you try to observe while the telescope is adjusting to the colder air you will see nothing but blurred images. Wait at least a half hour in the winter after sunset and fifteen minutes during the summer after sunset. The type of 'scope will determine how fast the optics reach a thermal equilibrium.

Never observe an object when it is over a rooftop. At night, especially in the winter, heat is rising from the rooftops of the nearby homes. This rising of warm air currents will cause a great loss of resolution in the object you are observing.

Also in the winter, never walk in front of the objective since the heat from your body will cause air currents which will distort the image you are observing. Make sure that if you are observing close to a heated home that no doors are opened which would cause a rushing out of heat.

Always plan your observing program before the night. You will find that things will work a great deal better if you have all your plans, calculations, notes, charts, and objectives all ready and worked out. Also, always have an alternative program in case the first one does not work out for some reason.

Be organized. Make an observing plan and stick to it.

When looking through your eyepiece don't put your eye right up to the lens. Keep your eye about a little less than a quarter of an inch away from the lens area. Moisture and oil from your eyelash can stain and fog the optics of the eyepiece.

Before you do any observing let your eyes get dark adapted. Deep-sky objects will show up much better if you look at them with your eyes adapted to the lighting condtions. Once you are dark adapted be careful that no bright lights are turned on that may glare in your eyes and ruin your night vision. To read charts use a flashlight with some red cellophane around it. The red light will not interfere with your night vision.

Do not watch TV before you observe. The light from the TV will "glare out" your eyes for sometime. You will also see images of the TV screen when you look into the telescope. These are after-images left on the retina of the eye and can blot out faint deep-sky objects.

In the winter, bundle up, for you may be out there for several hours. When using the telescope alwyays try to get in the most comfortable observing position. Do not constantly stare at an object through the eyepiece. If you do this you may end up with a headache before very long. Look at the object for about a minute at a time; this will give the eye relief that you will need.

Spend some time on each object that you observe. Try looking at it at different angles. Astronomers use a trick in observing deep-sky objects, called averted vision. While looking at a deep sky object such as a globular cluster look at it directly, then look at it from the corner of your eye. This can be done by looking directly at a nearby star in the field of view off to the side. You will notice that the cluster will look almost like a professional photograph with this type of observing. You will also notice that you will resolve much more and see more of the cluster using this averted vision then by looking directly at it. This is because the cells which are responsible for this side vision are more sensitive than the cells which give us straight-ahead vision.

Dew forming on the objective and the eyepiece can be a big problem. Dew forms as the telescope cools down and becomes slightly colder than the air around. This will cause a thin layer of moisture to form on the lens or mirrors. Dewing will cause a great reduction of image quality, and if it gets severe you might as well call it a night.

Reflectors of the Newtonian type do not suffer dewing unless there is a great deal of moisture in the air, as when you live close to a large body of water. This is because reflectors of the Newtonian type have their optics located deep in the tube. Also, refractors have their lens four to six inches down the tube. These are built in dew caps and they help prevent the building up of the moisture on the objective which causes fogging.

A dew cap is nothing more than an extension of the tube of the telescope, at least six inches from the objective. This extra tubing will help keep the objective warm to stop or delay dewing. Schmidt-Cassegrains do not come with any type of extra tubing. The corrector lens is located at just about the top end of the tube. Dewing is quite common with this type of system and although the objective may never fog up with moisture the corrector lens does. This fogging or dewing of the corrector lens is as bad as if the main objective itself had fogged. Dew caps also help keep extraneous light from entering the telescope.

You can buy dew caps for the Schmidt-Cassegrain systems and they are quite inexpensive. I suggest that you do invest in a dew dap if your system is going to need it. The amount of dewing that you get will depend on your area, time of year, and the type of telescope you use. Another way to beat the dew is to buy a heating system that hooks up around the telescope optics. This unit puts out a small amount of heat which will keep the temperature of the optics the same as the outside air.

There is no really good way to prevent dewing of the eyepieces, however, but they rarely fog up and will only do so on extremely wet nights.

The last point that should be covered here is care of your optics. All eyepieces and filters should be stored in a shock-proof, dust-free box. Your eyepieces are coated with special chemicals to reduce reflection. Do not clean them frequently. The best thing to do is to buy a can of compressed air similar to that used in the photographic darkroom and blow the loose dust off. These small air cans can be bought at any camera store. It will take quite a bit of dust on your eyepiece to interfere with light transmission. Oils on the lens are another story. They should be cleaned off as soon as possible since the oils will eat off the coating. To clean off the oil, you can take some Kodak lens cleaner on a cotton swab and gently make a small circular sweep of the surface. Do not wipe dry—blow dry with compressed air.

Cleaning your objective is another story. Refractors and reflectors will both collect dust. Many new amateurs will try to keep their optics dust free and clean from dirt. I knew of one young amateur who was so fanatical about keeping his reflector mirror clean that he used to wash it twice a day. This frequent cleaning resulted in the removal of his mirror's reflective coating. Not only did he put many scratches in his mirror but he also had to send it away to be recoated. It will take quite a bit of dust to interfere with the performance of your telescope so don't be concerned if you get a dust build up. Loose dust and dirt can be blown off the lens or mirror with a small can of compressed air. Your mirror or lens should be washed at least once a year and this is all. Refractors are much easier to clean since the lens optics are less likely to scratch. To do so, take a cotton swap and dip it into 70 percent rubbing alcohol and then, in a very easy smooth and light circular motion, make one sweep on the lens. Discard swab and repeat procedure until the entire lens has been covered. Take the dry air can and spray the entire surface of the lens in short bursts until the surface is dry.

The same procedure can be used when cleaning the corrector lens of a Schmidt-Cassegrain. The mirror of this system should never be touched since it is sealed inside the tube. If the mirror of your Schmidt-Cassegrain should ever need cleaning for any reason you must write the company that you bought it from for further instructions.

To clean the surface of a Newtonian reflector the same materials are needed, plus a screwdriver to remove the telescope mirror. Since in most cases the objectives of the Newtonian mirrors are larger than their refractor counterparts you will have more surface to clean. Also the aluminized coating on your mirror is very delicate and can be scratched very easily. Another danger is that in this case you have to take the mirror out of the tube to clean it. If you have an eight- or ten-inch or larger it could be very touchy since if you drop it or drop something on the surface it will probably chip or crack.

Carefully take the mirror out of the tube with its cell or mirror mount. Place on the table, and loosen the mirror clamps on all sides. Using the surgical cotton roll, pour some alcohol on the cotton and gently wipe the mirror softly in a circular motion. Discard cotton after one complete sweep of the mirror and repeat until all the mirror surface has been covered. Then take your can of dry air and blow the mirror dry. You may see some slight discoloration if you hold the mirror at an angle to a light but this is all right.

Your mirror will be clean now. Put it back into your telescope tube and don't take it out to clean it again at least for another year.

When you put your telescope mirror back into the tube you will have to realign it. This information you will find in your telescope instruction booklet. It should also be noted that your elliptical secondary mirror can also be taken out and cleaned in the same way.

The coating on your mirror's surface will wear with time. The amount of time your mirror's reflective coating will last will depend upon your location, how well you care for your optics, and the quality of the coating put on. I had one ten-inch mirror some years ago that, when I cleaned it, the coating washed right off. This should not happen; the coating on this mirror was of very poor quality, and so was the telescope. I won't mention the company since my case may have been an isolated one and the company is well known and has been making telescopes for years.

The coating on your mirror should be redone every five to

ten years. There are all different types of coatings and the standard-priced jobs will give you 85 percent reflectability. The higher priced ones will give anywhere from 93–99 percent reflectability, the price going higher and higher with the percentage. For example, a coating with 93 percent reflectablity will run about $25.00 for an eight-inch mirror. A 99 percent coating will run over $150.00. The human eye can not discern the difference in reflectability. In short it is a gimmick and not worth the price for that extra 6–7 percent. If you have to recoat, stick with the coatings that give you at least 90 percent. This will give you a good quality coating and high reflectability.

Companies that do recoatings are listed at the end of this book.

IX

ASTRONOMICAL CLUBS AND ORGANIZATIONS

There are quite a few astronomical organizations in the United States and chances are there is one right in your area. Though these local groups provide no major project work or research, they do offer meetings once or twice a month, lectures, and perhaps occasional use of a club telescope.

It could be a very good idea if you would check you local area for astronomy groups. This information can be obtained from your telephone book or local library. Through these groups you will meet experienced amateurs and professional astronomers and astrophotographers and it is a good way to get started.

You should also join one of the several nationwide amateur–professional groups which offer to let the amateur help in scientific research. I have listed the major nationwide organizations below and will give a brief description of what they do and how they operate.

The Association of Lunar and Planetary Observers (ALPO) Box 3SZ University Park, Las Cruces, New Mexico 88003.

ALPO is an informal international group of observers who observe and study the moon, planets, and comets. The organization's goals are to stimulate study of these bodies by amateur astronomers. ALPO provides a service for the advanced astronomer as well as the novice astronomer who wishes to use his talents and equipment for the study of certain objects. All interested people are welcome as members and since all work is done on a volunteer basis each member can do as much or little as he or she wishes.

The work of ALPO is coordinated by means of a periodical, *The Strolling Astronomer*, published four times a year. Observational activities are divided into groups. *Mercury section* provides observational study of the planet. *Venus section* records various phenomena in the cloud cover. *Mars section* studies the polar caps, surface changes, and dust storms. *Jupiter section* observes the Jovian cloud system, storms, red spots, and so on. *Saturn section* studies the ring system, Saturnian moons, cloud belts, and storms. The *remote planets section* has a number of projects including studies of Neptune, Uranus, and Pluto. The *minor planets section* is working on a complete study of the asteroids. The *comet section* is making systematic observations to discover new comets. Studies also include observing known returning comets.

Of course the study sectons are more involved than indicated above. ALPO is a fine organization that provides serious astronomical research. If you are interested I suggest you contact Professor Walter H. Haas at the organization's address.

The American Association of Variable Star Observers (AAVSO), 187 Concord Avenue, Cambridge, Mass., 02138.

The AAVSO was founded in 1911 and is a not-for-profit organization of professional and amateur scientists whose goal is to study variable stars. AAVSO is a worthwhile organization with 1,100 members in 38 countries. The programs include studies of Cepheid variables, eclipsing binary stars, RR Lyrae stars, solar research, nova search, photoelectric photometry studies, and the publishing of their newsletter and other materials. All work of the members is reported in two journals. *AAVSO REPORTS* (contains observations), and the *Journal of the AAVSO* (contains organization information, report of the director, results of projects, and so on.

AAVSO is a semi-professional organization which requires expert training in the field of variable-star research. Advanced amateurs are welcomed to inquire. Novice astronomers are also welcome to write for information and training guides which will lead to future membership.

If you are interested in a study of the stars I suggest that you contact AAVSO for further information. Membership is somewhat restricted but novice astronomers are accepted occasionally.

The Astronomical Society of Greater Hartford. This is more or less a local organization for the Connecticut area. It provides use of a club telescope and meeting and lectures. I have done some research of this organization and it is a fine group of advanced and novice amateur astronomers who have very interesting programs. If you live in the Hartford, Conn. area contact them at 17 Greystone Road, West Hartford, Conn., 06107.

The American Meteor Society (AMS), 32 Woodland Court, Route 3, Callahan, Fla. 32011.

The AMS was founded in 1911 by Dr. C. P. Oliver and is made up of both professional and amateur astronomers who are interested in meteoric study. Members submit records of observations of meteor showers and fireballs. While associate membership is open to all over age 18 you must show that you have a great interest in the work being undertaken. AMS also accepts novice members aged 14 and over. If you have an interest in the study and observations of meteors you may contact their headquarters at the Department of Astronomy of the State University of New York at Geneseo, N.Y.

The National Observers of Aerial Phenomena, P.O. Box 4218, Greenwich, Conn. NOAP is an organization which brings together professional and amateur scientists who wish to document various aerial phenomena.

Conceived in the late 1960s, NOAP grew into a nationwide organization in the early 1970s. Although it may appear to be a typical UFO investigation agency from its name, NOAP interprets "aerial phenomena" as encompassing astronomical, meteorlogical, and UFO phenonena.

NOAP carries out projects on various subjects which it asks its members to to contribute to. Studies undertaken will vary depending on which division you belong to. For example, the astronomical division studies phenomena on the moon, sun, planets and some deep sky observation. The meteorological divsion studies and records interesting meteorological phenomema, storms, sky glow, rainbows, and so on. The UFO division works in cooperation with The Center for UFO Studies and is interested in gathering UFO reports for further investigations. For further information contact NOAP at the address above.

APPENDIX: Manufacturers' Listing of Equipment

Below is a listing of manufacturers of high quality astronomical equipment. I have listed only those companies with which I have done business—those that have not only very good equipment but also good service. Over my fifteen-plus years in astronomy, both as an amateur and professional, I have done business with quite a few companies and like everything else in this world there is the good and the bad. The list below contains not only names of the companies but also a sublisting of their best and most popular equipment.

Rocky Mountain Science Supply, 3093 Walnut St., Boulder, CO 80301.

Refractors, and reflectors from three to twelve inches. All types of compound telescopes, three to eight inches. Large supplier of all types of astronomical accessories. Catalog: $2.00

Kenneth Novak & Co., Box 69 Ladysmith, WI 54848.

All good quality telescope parts for construction of Newtonian reflectors. Catalog: Free

MMI Corp., 2950 Wyman Parkway, Baltimore, MD 21211.

Astronomical slides, poster, and tapes; Star maps, books, and educational material. Some astronomical equipment, also. Catalog: Free

Questar Inc., Box Q12, New Hope, PA 18938.

Compound telescopes (Maksutov type) three to twelve inches. Expensive, but high quality. Catalog: Free

Coulter Optical Co., Box K, 54121 Pinecrest Rd., Idyllwold, CA 92349.

Telescope kits from four to twelve inches. Catalog: Free

Celeston International, 2835 Columbian St., Box 3578, Torrance, CA 90503.

Schmidt-Cassegrain telescopes from eight to fourteen inches. Catalog: $2.00

Unique Scientific Systems Inc., 1130 51st Street, Brooklyn, NY 11219.

Eyepieces, nebula filters, finder telescopes, dew caps, variable-speed controls, telescope parts, and accessories of all types. Catalog: Free

The Astronomical Unit, 3704 John F. Kennedy Blvd., North Little Rock, AR 72116.

Refractors two to four inches; reflectors six to twelve inches; Schmidt-Cassegrains eight to fourteen inches. Astro cameras, filters, lenses, tripods, astro film (spectroscopic 103 series and SO-115), eyepieces, finders, guide scopes, illuminated cross-hair recticles, and variable speed controls. Catalog: $2.00

Meade Instrument Corp., 721 West 16th Street, Costa Mesa, CA 92627.

Refractors 2.4–4 inches; reflectors, six to twelve inches. All types of quality eyepieces, complete telescope guiding system for deep sky photography, high-quality nebula filters, astronomical filters of all wavelengths. Catalog: $2.00

Rudy E. Kokick Co., 33-05 Murray Street, Flushing, NY 11354.

Cold camera for deep-sky work, Special astrocameras with special vibration-free shutters, high quality fast telephoto lenses for deep sky work, off-axis guiders for compound telescopes, guide 'scopes for refractors, reflectors, telecompressors for compound telescopes (changes f/10 to f/5); other high quality telescope parts, and scientific equipment. Catalog: Free

Edward R. Byers, 29001 West Highway, Barstow, CA 92311.

Very high-quality clock drive systems and telescope mounts. Catalog: Free

Unitron Scientific Company, 101 Crossways Park West, Woodbury, NY 11797.

High-quality refractor telescopes from two to six inches.

Roger Tuthill, Inc., Box 1086 ST, 11 Tanglewood Lane, Mountainside, N.J., 07092.

Maker of Solar Skreen full-aperture filter for solar observing; telescopes of all sizes, top-quality astronomical equipment. Catalog: Free.

Astronomy Magazine. 411 E. Mason St. P.O. Box 92788, Milwaukee, WI 53202.

Sky and Telescope. 49-50-51 Bay State Road. Cambridge. Mass. 02138.

Dudley Clausing Company, 8038 Monticello Ave., Skokie, Ill., 60076.

High Quality 93% Reflective Beral Mirror Coatings.